The Sequential Imperative

AF714599

Value Inquiry Book Series

Founding Editor

Robert Ginsberg

Executive Editor

Leonidas Donskis†

Managing Editor

J.D. Mininger

VOLUME 301

Cognitive Science

Edited by

Francesc Forn i Argimon

The titles published in this series are listed at *brill.com/cosc*

The Sequential Imperative

General Cognitive Principles and the Structure of Behaviour

By

William H. Edmondson

BRILL
RODOPI

LEIDEN | BOSTON

Cover illustration: The photograph (taken by the author) is of part of the exterior wall of a long, low building on Suomenlinna, a maritime fortress built on a group of islands near Helsinki. Part of the building houses the library for the community.

Library of Congress Cataloging-in-Publication Data

Names: Edmondson, William, author.
Title: The sequential imperative : general cognitive principles and the structure of behaviour / by William H. Edmondson.
Description: Leiden ; Boston : Brill-Rodopi, 2017. | Series: Value inquiry book series, ISSN 0929-8436 ; VOLUME 301. Cognitive Science | Includes bibliographical references and index.
Identifiers: LCCN 2017022617 (print) | LCCN 2017031793 (ebook) | ISBN 9789004342996 (E-book) | ISBN 9789004342897 (pbk. : alk. paper)
Subjects: LCSH: Cognition. | Cognitive science. | Linguistics. | Human behavior.
Classification: LCC BF311 (ebook) | LCC BF311 .E329 2017 (print) | DDC 153--dc23
LC record available at https://lccn.loc.gov/2017022617

Typeface for the Latin, Greek, and Cyrillic scripts: "Brill". See and download: brill.com/brill-typeface.

ISSN 0929-8436
ISBN 978-90-04-34289-7 (paperback)
ISBN 978-90-04-34299-6 (e-book)

Copyright 2017 by Koninklijke Brill NV, Leiden, The Netherlands.
Koninklijke Brill NV incorporates the imprints Brill, Brill Hes & De Graaf, Brill Nijhoff, Brill Rodopi and Hotei Publishing.
All rights reserved. No part of this publication may be reproduced, translated, stored in a retrieval system, or transmitted in any form or by any means, electronic, mechanical, photocopying, recording or otherwise, without prior written permission from the publisher.
Authorization to photocopy items for internal or personal use is granted by Koninklijke Brill NV provided that the appropriate fees are paid directly to The Copyright Clearance Center, 222 Rosewood Drive, Suite 910, Danvers, MA 01923, USA. Fees are subject to change.

This book is printed on acid-free paper and produced in a sustainable manner.

Contents

Preface

This book is unusual. It is about a Big Idea – the Sequential Imperative – which appears not to have surfaced before. The Big Idea – stated here in simplified form – is just that the brain manages behaviour and perception by providing its host with the mechanism for dealing with the *unavoidable* need to transform the constantly shifting and changing "external world" into the timeless, or atemporal, "internal world" of memories, perceptions and so forth. And at the *same time* the brain takes plans and intentions – "internal world" atemporal mental entities – and transforms them into sequences of activity, including language, in the "external world". The brain serves as a curious sort of bi-directional time machine – and this book explores the details of why and how it does this, all without getting into biological detail!

The book appears to have a diffuse focus: it is primarily about Cognitive Science but it is also about speech and language; it ends up fusing these two fields. We'll review this throughout the book. It is a monograph which resembles a series of notes for bigger projects; it is not a work of reference. This is discussed shortly. The note-form encourages a quirky style, and we'll return to this point later.

Let me explain. The people involved in Cognitive Science include Linguists, Psychologists, Philosophers, Neuroscientists, Computer Scientists, and Cognitive Scientists (who are themselves licensed interdisciplinarians). Cognitive Linguistics has been identified in recent years as the interdisciplinary study of human language behaviour, involving the same range of disciplines. The separate identification of Cognitive Linguistics as a topic, alongside Cognitive Science, can be thought of as a consequence of recognizing the richness and complexity of human language behaviour. Linguistics is recognized as a separate discipline in academia but doesn't have the Cognitive Science component in its motivation. So Cognitive Linguistics is an attempt to bring the isolated study of language behaviour closer to where it should be: part of the study of cognition generally. This book breaks new ground by offering a unified interdisciplinary approach to studying both cognition and human language behaviour. In doing so it effectively leaves Cognitive Linguistics without its own domain of study and thus without motivation. However, before we get to that point it is useful to look at what motivates Cognitive Linguists currently.

The motivation for Cognitive Linguists is the search for explanations of language behaviour which are not grounded in uniqueness. Of course, humans are the only species to use human language (there are thousands of such languages) but Cognitive Linguists consider that the uniqueness of *language* is actually no different in principle from other human uniquenesses (music, graphic art, etc.) as well as the uniquenesses which define any other species. We see unique activity in humans, to be sure, but should that imply an array of unique mechanisms and processes unrelated to anything else in humans or other animals?

The Cognitive Scientist is challenged by this perspective – surely they should be able to explain how languages work so why is it so difficult? This book responds to that challenge by showing instead how it is possible to develop ideas derived from the study of speech and apply them to Cognitive Science generally. Indeed it goes further by explaining how accounts of locomotion, control of articulators, perception, learning, and so forth can be built on detailed understanding of how speech is produced and perceived. I further argue that such accounts are applicable to the study of all organisms with brains, not just humans, and also that such accounts can provide for the possibility of language in a species such as ours, but not necessarily in all species. Cognitive Science must account for the possibility of language behaviour – this is a theme in the book.

As part of the work of explaining how Cognitive Science provides an account of the possibility of language, the Cognitive Scientist must also address the evolutionary question: How can any brain come to support language behaviour? In this book we find that the general Cognitive Science account of behavioural possibility (including evolutionary concerns) can be derived from our understanding of speech behaviour in humans. This is surprising until we recognize that speech is the most intensely studied of all animal behaviour.

It should be noted that some linguists hold the view that human brains are somehow specified for language with an innate endowment (Universal Grammar) that makes language possible – an endowment which is unrelated to more general cognitive principles. This book argues in favour of an alternative view – that general cognitive principles can be found to explain the possibility of language behaviour. The book is not motivated as an attack on Universal Grammar but is rather an attempt to provide the bigger Cognitive Science account required to encompass language behaviour.

The content thus deals with two related questions – one looking at the challenge "from the top down" so to speak, the other looking "from the bottom up". The questions are: How do we do Cognitive Science in such a way that the possibility of language behaviour is addressed inevitably? How can study of speech in humans contribute to the development of a Cognitive Science which covers non-human species? I propose that within a general Cognitive Science "top-down" context the inevitable tight coupling between the two endeavours is readily achieved when we identify from the "bottom-up" some cognitive generalities (for all species) through the study of human language behaviour. This is a difficult trick to pull off, and in any case the more general "top-down" account will need to come first in any exposition, so the assembly in the reader's mind of the unified account requires some material to be taken on trust until the rest is offered.

The book is suitable for the informed general reader, for advanced students and active researchers alike. There is a possible problem for the reader, however. The Cognitive Science and Linguistics core material is readily accessible for those working in these domains, but the material is interdisciplinary, and some readers may struggle with some of the range of topics covered – from phonology to astrobiology. In addition, it is my view that the book is paradigm shifting – some readers may find that difficult.

Turning now to the quirkiness of style, let me re-iterate that the book is explicitly a prolegomena – a set of notes for bigger projects to come. It is a monograph not a textbook; it is broad in scope but without a monumental list of references or doorstop heftiness. The nature of notes is incompleteness and, to switch to a graphic metaphor, the set of notes amounts to a sketch which can do much to reveal the overall picture, along with the size of the canvas, without covering it all. Some material will be well detailed, some in outline awaiting adumbration, and so forth. But all combine to depict a totality in the image.

However, the quirkiness does not stop with the sense of sketchy notes. The core material is presented as a set of numbered paragraphs. This discourages verbosity and permits rapid changes of topic and focus, but creates a feeling of staccato terseness and density that may be disconcerting. The book is actually quite short – around 80,000 words. The numbered paragraphs are not found in the *Introduction* which flows in more narrative style.

The function of a *Preface* in a book is to set out the stall for the reader; the browser in the bookshop should by now know whether or not the book is "for them". The *Introduction* which comes next sets out the scope of the material on the assumption that the reader is comfortable with the idea of trying to understand questions such as "What is the brain for?".

A *Preface* is also the place for acknowledgements and thanks. My career, when viewed retrospectively, has been non-linear in academic terms. Following a degree in Physics I spent some years working for a doctorate in a combination of Electrical Engineering and Psychology. This was followed by some observational studies of deaf children in educational settings, and an increasing involvement during the late 1970s in the small UK research community working on Signed Language Linguistics. After a short break I consolidated my interest in Linguistics with a Masters degree in Theoretical Linguistics whilst at the same time developing an active interest in the study of human behaviour with computers – a field known as Human-Computer Interaction (HCI). I spent nearly 25 years in employment at the School of Computer Science at the University of Birmingham teaching and researching in HCI, Speech Processing, Signed Language Linguistics, and Cognitive Science generally. Towards the end of my time at Birmingham I added work on Astrobiology and SETI to my research portfolio. At every turn there have been people offering encouragement and support – too numerous to mention here, but vital to me for providing an environment where I could see how the pieces linked together. No-one is to blame for my winding intellectual journey except me. Although the ideas gathered along the way have fused together to form the Big Idea presented here the reader should rest assured that the presentation is not a travelogue. The book is coherently organized to present the content thematically.

My wife Linda has been a huge support during life's journey and the struggle to get the material of the book actually into book form. Some people have read earlier drafts and made some helpful comments – Paul Brinich, Matt Colborn, Peter Coxhead – but they are not to blame for any errors. An anonymous reviewer for Brill also made extensive really helpful comments. Brill's editorial team, from manuscript review through to production, have been brilliant: many thanks then to Francesc Forn i Agrimon, Bram Oudenampsen, Thalien Colenbrander, Gera van Bedaf, and Kim Fiona Plas.

List of Figures

Introduction

This book is about the *functional specification* of the brain. More than four decades spent working as a Cognitive Scientist, focussing primarily on human communication, but active also in the field known as Human-Computer Interaction, provide the experiential basis for the conjectures set out here. Although I have worked on some aspects of *human* behaviour – spoken and signed language linguistics, theoretical work in cognitive ergonomics – my conjecturing concerns behaviour more widely, and in any creature with a brain. A career's worth of work and observation naturally encourages me to seek some generalisations but those offered here are so sweeping that the outcome might amount to little more than self-centred and grandiose intellectual over-reaching. That remains to be seen. Put simply, I have come to see the need for an answer to the question: *What does the brain do?*; or to put it more colloquially: *What is the brain for?*. This book sketches out an answer, and it does so by explaining what I consider to be the functional specification of the brain. Notice immediately that the focus is not the human brain, but the brain more generally. As noted in the Preface, the answer we explore here is that the brain serves as a curious sort of bi-directional time machine. We will state that more formally in Chapter 1. The exploration is presented in the remainder of the book.

Because the notion of *functional specification* may be a bit obscure it is useful to look at a straightforward example from technology (the term is used in software engineering, but is valid more widely). A time-switch – such as one found in many homes for controlling lighting, or central heating, or whatever – controls the electric power to one or more devices/ appliances at times set in advance by the user. That is what it is for. When asked by someone who has never encountered such a thing, the user/ owner will describe what it does – they will actually provide a *functional specification*. They will explain that the user can control the pattern of periods of power supply (either "on" or "off") which may be variable over a week, or more typically over a 24 hour day. The functional specification is simply the behaviour offered to the user for control – how many periods of "on/off" during a day? Are weekends special? Can the user over-ride a setting and if so will the device eventually revert to the preset pattern or does that have to be controlled separately? And so forth.

Many people have encountered such time-switches. Their *functional specification* is usually set out in an instruction book, but generic characteristics can be carried over to a new device based on experience with other devices. What is not relevant is *how* the functional specification is delivered – indeed the user/ owner may not know and does not need to care. The mechanism

© KONINKLIJKE BRILL NV, LEIDEN, 2017 | DOI 10.1163/9789004342996_002

could be brass, plastic, electronic, old beer cans, whatever; it really doesn't matter. Occasionally there will be cross-coupling between the technology and the functionality. What is the smallest increment in time that can be used to determine the settings? With electronic timers the control is probably specified in minutes. For devices with a rotary mechanism it is more likely to be quarter hours or hours. But this detail doesn't impinge on the functionality as such. Thus, when considering the functional specification of the brain we do not need to say anything about the wetware that neurosurgeons get to see and neuroscientists get to measure and image (there is some brief coverage in Chapter 2 for reasons explained later). Indeed modelling of the wetware in computers is also irrelevant. The computer modelling we will look at briefly (Chapter 6) deals with the structure of behaviour and is conceptual. It is not a computational model of the brain, it is a computational model of the structure of behaviour and of how the brain's functional specification delivers behaviour.

My work in Cognitive Science generally is distilled in the conception of a *functional specification of the brain*. This book sets out the conception for the reader. For me the work also reflects an attitude to the study of human language behaviour which takes on board the complexity of linguistic structure and activity but does not seek special explanations through appeals to uniqueness of cognitive endowment in the species. In other words I seek to understand speech and language as part of general cognitive capabilities.

In direct contrast, Universal Grammar (UG) projects the need for a special cognitive – indeed neurological – endowment as the basis for explaining language behaviour in humans. Linguists, and also Psychologists and Cognitive Scientists, vary in the degree to which they cleave to the details of UG but many find the sense of uniqueness of human language compelling and thus UG attractive in a general sense. The problem with UG, in relation to the material in this book, is that it is inadequate as a basis for a general account in Cognitive Science (irrespective of its value as a linguistic theory). Specifically, the account on offer here is couched in terms of General Cognitive Principles (considered by Chomsky to be potentially undermining of UG, as we will see later), and goes on to address issues in learning and evolution in ways which critique UG. The critique is not the motivation here, it is just part of the larger unfolding account. Indeed the material of the book covers concepts quite outside the remit of UG and thus contextualizes what needs to be said about language learning and the evolution of humans as language using primates. I dwell on this here to reassure two groups of readers. Those who consider UG to be significant for Cognitive Science should not be dismissive of something they think will emerge as yet another attack on UG; those who think UG is not relevant to Cognitive Science should not set the book aside because they think it will say nothing new for them. I go further and challenge both groups of readers, as well as other readers and those working in Cognitive Science. My claim is that understanding speech and language takes us beyond the uniquely human in behaviour. It is the key to progress in reconceiving Cognitive Science as a pan-species endeavour.

I assume there is no need for an endowment for language which is unrelated to other cognitive activity or potential, but I must then offer a coherent alternative account of how language can exist and work. There is thus a difficulty with a book such as this one – how to avoid appearing to be two books rolled into one; one covering Cognitive Science, the other dealing with Linguistics. The failure of Cognitive Science to account for language behaviour so far, along with the failure of Linguistics to develop any comprehensive theory which fits with Cognitive Science, is resolved by showing that Cognitive Science itself can be developed further on the basis of insights from the study of language behaviour. Speech is the most intensively studied behaviour in any animal and it is surprising that Cognitive Scientists have not recognized the relevance of the details.

Cognitive Science must account for the possibility of language behaviour – this is a major theme of the book. Our concern must be the development of ideas in Cognitive Science showing how the study of speech is important for more general understanding. Rather than trying to demonstrate and analyze inadequacies in work based on current theoretical preoccupations, my approach is to start at the end and work back. In this way I will try to reveal (rather than argue) that Cognitive Science has reached the point where we need to step back a little and reconsider what we can do if we understand speech a little better, and if we work with a more general view of what the brain does. This may seem lazy, but the motivation is simply that the functional specification of the brain is *not* derived but is instead distilled from observation. One potential disadvantage of this approach is that modern developments in some of the sub-disciplines of Cognitive Science (especially neuroscience) are ignored – and this might seem incompetent. As explained above – exploration/ discussion of functionality does not require exploration/ discussion of the biology of the delivery of that functionality.

It is possible that what the reader finds here will be stimulating and provocative in the manner of, say, *Plans and the structure of behavior*[1] – an influential book in its day, but now not much used. So maybe the only effect of work on the functional specification of the brain will be that in 50 years time historians

1 Miller, G.A., Galanter, E., and Pribram, K.H. *Plans and the Structure of Behavior.* Holt, Rinehart and Winston, 1960.

of science will be able to discern patterns of influence which fall somewhat short of pervasive or revolutionary (and it can be argued that the development of cybernetics as a model for the functionality of the brain was another failed effort to understand behaviour[2]). Perhaps, alternatively, the work presented here could suffer in the way that Wegener's conception of *continental drift* was treated – it was considered in his time to be speculative and insufficiently supported by evidence. Another precursor, this time in terms of failed grandiosity, could be Haeckel's *biogenetic law* – summarized as "Ontogeny recapitulates Phylogeny". Haeckel was an exquisite draftsman[3] and his law is now just an appealing but simplistic error of judgement, although sometimes valued in explanations (we will encounter later some instances of this in relation to the origin of speech).

So what is going on here? Your author is pre-occupied with the long-term value of his ideas almost before the work is started! Like any author, I consider that this book's value will endure longer than the time taken to print it; but there are other issues. Some of the ideas which follow are going to challenge many readers because they derive from rather abstruse work in phonology (the linguistic study of systems of articulation deployed in languages) which I think it is necessary to appreciate if the bigger picture is successfully to emerge. Which is to say, the value of the ideas may be undermined by the route taken to expose them. It may of course be the case that once the overall conception is "out there" and the functional specification understood, then others with better expositional skills will find easier ways to tell the story. I am, in one sense, hampered by "going first". Although I have carefully avoided a diary-style account of how I came to have the overall idea, with all its detail, I am nonetheless prisoner of my own conceptual process. The Big Idea is not, of course, a property of my way of thinking, but it might seem so on its first big outing (see also Edmondson, 1986a, 2010). In consequence I acknowledge that although it is perhaps unfashionable to expect the reader to "work at it" I think it is pretty much unavoidable. Of course, the ideas will be presented in various ways, and with different models or analogies to help the reader. But ultimately there is work to be done by the reader as well as the author.

So, why is a functional specification of the brain key to understanding cognition and language? What is to be offered the reader here so that they turn the pages wanting to put in some effort in order to know more? Questions, always questions – but to repeat, and indeed elaborate: What is the brain for? What

2 This point is taken up in Chapter 6; see also Frankish and Ramsey, 2012.

3 See Olaf Breidbach's beautiful book about Haeckel: *Visions of Nature: The art and science of Ernst Haeckel.* Translated by Paul Aston. Prestel, 2006.

does the brain do (for its host)? What is the commonality between the brains of humans, chimps, cats, dogs, crows, chickens and all the rest? How can we build a general account of behaviour in all species on our understanding of human language? A functional specification will help us answer these questions. It will also shape the future of Cognitive Science.

It is useful here to point out that we have some simple notions for describing organs in a pan-species way. We can recognize that the heart as an organ is unhelpfully described as "the organ that keeps you alive". In contrast, the functional specification of the heart as the body's pump for its blood supply and circulation system seems about right. The precise details – such as the two circulations (systemic and pulmonary) pumped by the single organ – are not actually relevant in general functional terms. And the brain? What is the function of the brain? What does it *do*?

The conventional Cognitive Science answer to this question is to say that the brain is the organism's "information processor". Indeed, much work in Cognitive Science and Artificial Intelligence assumes this answer and develops information processing schemes, theories, models, etc. on the basis that these inform us about what brains do. Computational models of the wetware are of course dependent on the latest research into how the wetware works, but are nonetheless focussed on the wetware as computationally active (hence the term wetware). Actually, it is not clear that this work is much more than redescription of the problem. One difficulty is that there seem to be unlimited numbers of models, schemes, theories available for the work – so the apparent answer to the question is never finalisable. Another difficulty is that information processing is inadequately defined and lacks the self-evident validity that "pumping blood" has when talking about the heart. For example, I can point out that televisions, telephone exchanges and portable computers are all information processors, but one learns nothing about the functionality of these devices from such a general description.

And just in case the reader is impatient for an answer – why is it important to find a functional specification of the brain? – my point is really that without it Cognitive Science has, in some sense, lost its way. There has been no account (until now) of what the brain *does* so progress has stalled. For example, Linguistics is not really treated *within* Cognitive Science and this has given rise to the sub-discipline of Cognitive Linguistics – but if Cognitive Science were more mature and more comprehensive there would be no motivation to invent Cognitive Linguistics (cf. Ch.9 of Frankish and Ramsey, 2012, where Jackendoff opens with: "Within cognitive science, language is often set apart (as it is in the present volume) from perception, action, learning, memory, concepts, and reasoning".). There has been a tendency to work on parts of the system at the expense of the whole; to work on how bits of the brain could (or could not) work, rather than on what the whole thing is for; to work with rather atypical behaviour (because human) rather than to consider pan-species generalities.

As noted above we will avoid deep analysis and dissection of failures and inadequacies in existing work in Cognitive Science. Instead we will start at the end and work back. In reality this is a tidying up of the emergence of the ideas in my work, because there is no reason to subject you to the meandering path I took to get here. One point, which will be addressed later, is that in the early 1980s, when working on sign language morphophonemics (looking at the way different components of meaning can be part of the overall deployment of the hands in signing), I had an unusual idea for investigating the then popular conception that signing is very imagistic (with the meaning components being much to do with depiction). This work – with a comment from a colleague – started a particular train of thought which ends up with this book. More on that idea later (see 3(8).16.2). Also important is that I frequently refer to work published in the last century. There is a sense in which Cognitive Science could have "gone a different way" had it applied different emphasis to some of the popular work in the 1960s–1990s, and paid attention to some of the even earlier work in Linguistics. I think it is useful to understand the missed opportunities. But, as I said, this book is not a review of the history of Cognitive Science (cf. Abrahamsen & Bechtel, 2012). Early work will be cited where it seems especially pertinent.

In the first chapter we encounter the core idea – the Sequential Imperative. This is the Big Idea which is boldly simple and which shapes the Cognitive Science in this book. The first three chapters illustrate and explain the concept in non-technical language, taking us about one third of the way through the book; this is the top-down perspective of Cognitive Science (Part 1). And because we know where we are going in this account the Cognitive Science is set out in such a way as to meet (3 chapters in Part 3) the bottom-up account from the study of speech and language (3 chapters in Part 2).

The mid-third of the book (Part 2) is more technical and provides a Linguistics perspective – from the bottom up. Examples are drawn from linguistic behaviour but are linked to non-linguistic behaviour in humans and non-humans. A core idea is a scale-independent pan-species model derived from theoretical work in phonology. The remainder of the book (Part 3) fuses the two perspectives and elaborates the ideas, explaining what is required to actually make the Sequential Imperative deliver results, via the brain, for an organism. The material also covers evolutionary and philosophical implications, and – perhaps inevitably – offers a functional solution to the mind-body problem.

How should you approach the material here? It is very wide-ranging in scope – so you could perhaps "scope it out", as they say, much as you might quickly walk through an art exhibition to determine the range and quantity of the material on which you will dwell at a slower pace. You could try this by reading through the introductory material for each of the three Parts. On the other hand, such "getting the general picture" is much more difficult to do with text, so perhaps you should just get stuck in.

Those who find the prospect of tackling phonetic and phonological details daunting might want to take the second Part, especially Chapter 5, slowly and refer to some text-books. However, I have tried to provide a gentle lead-in to the material so that the details and complexity can be appreciated without reference to other books. The subtlety and variety of structural detail really does illuminate the bigger picture so it is really worth persevering. Perhaps most bizarrely of all, in the final Part some space will be given to considering what we can say about brains "off-planet" – brains of Aliens. Is it possible to conclude anything about such brains, and (starting with ideas derived from human speech, remember) decide whether or not they must conform to the functional specification of brains on Earth? Read on and maybe by the time you get to that material you will have formed your own view.

PART 1

The Sequential Imperative and the Functional Specification of the Brain

Introduction to Part 1

In this Part, comprising three chapters, we look mostly at the organisation of human behaviour other than speech and language, as well as non-human behaviour. We consider some general – anecdotal, even – accounts of how behaviour is patterned in sequences. We also see how the details of sequences illuminate the reality of sequentiality – sometimes arbitrary, sometimes specified (Chapter 1). We learn that the details matter at some scales of description and that arbitrariness in behaviour doesn't block a sense of purpose or the attainment of goals. We also acknowledge, but do not explore in detail, the issue of homeostasis – the preservation of the state of being of an organism. The brain plays a rôle in this, but we can set the issue aside until later in the book. The second chapter looks at how activity in the brain is patterned in time, on timescales varying from tens of microseconds to seconds, and at the rather limited extent to which any of this activity matches the pattern of activity in the "external world". The point is to show that the internal workings, so to speak, are not relevant to the functional specification (as with the time-switches discussed earlier) and that this was knowable in the middle of the 20th century. The third chapter looks at some general cognitive principles which need to be addressed if the brain's functional specification is to be made to work coherently. We start now with a statement of the functional specification of the brain – any brain, anywhere, at any time, in any species.

1(0) *The functional specification of the brain is to serve the sequential imperative.* We begin therefore by considering the Sequential Imperative. We return later to consider the complexities of how the brain serves the Sequential Imperative.

It turns out, unsurprisingly perhaps, that the devil is in the detail. The brain has quite a lot to do in the service of the Sequential Imperative, and some of what it does is quite complex. Some of this complex material is presented briefly in Chapter 3, to be taken up later in the third Part of the book. But ultimately, and for ease of tackling all the complexities offered in this book, the reader should keep in mind the first paragraph of the *Introduction* and the comment made there: *The brain serves as a curious sort of bi-directional time machine.*

© KONINKLIJKE BRILL NV, LEIDEN, 2017 | DOI 10.1163/9789004342996_003

CHAPTER 1

The Sequential Imperative – I

1(1).1 The Sequential Imperative (SI) is the constraint, or requirement, that any creature must pattern or modulate its activity over time – that is what constitutes behaviour and differentiates it from mere existence. Maintenance of existence as an organism, (homeostasis) is about maintaining the integrity of the organism, not about lack of activity. The SI is the requirement that a creature has to be active, but that means inevitably that some actions will follow others. Sequentiality is thus an inevitable part of an organism's active existence and engagement with the world. We will find a more complete account of how the Sequential Imperative develops as the story unfolds, but for now the focus on sequential activity serves us well. For the remainder of this book we restrict our attention to consideration of organisms with brains. Further, we are not concerned in detail with the brain's rôle in homeostasis although clearly the Sequential Imperative cannot be avoided – there is necessarily some sort of representation of "desired" body state, and of any "deviation" from that state, and therefore of what to do about correcting the "deviation". This is briefly discussed later.[1]

The alternative to sequentiality, of course, is simultaneity. However, it makes no sense to imagine an organism could simultaneously do everything in life it has to do, but in one short burst of activity. And there is the fact that simultaneous doesn't mean instantaneous – so sequential activities would be observable even as the organism expressed its totality in an unbelievable burst of simultaneous feeding, excreting, growing, breeding, moving, not moving, and dying.

No, even where simultaneity *is* indeed a part of our story, as when muscle fibres act in co-ordination to achieve an outcome, or articulatory activity involves groups of muscles in co-ordinated contraction and stretching to achieve an outcome, such simultaneity is better understood as sequential patterning of co-ordination. The interesting thing about sequentiality is that although different types of sequential arrangement can be discerned – as we will see shortly – the rôle of the brain in managing the sequentiality does not change.

1 See 1(1).4.9.1, 1(3).7.9.1 and 3(7).13.6.3. The point here is not that homeostasis is uninteresting, or that it cannot involve cognition and behaviour – think of shivering. Our primary concern in this book is a Cognitive Science account of behaviour and language. See also Chapter 6 and the discussion there of cybernetics.

© KONINKLIJKE BRILL NV, LEIDEN, 2017 | DOI 10.1163/9789004342996_004

1(1).1.1 Let us look at sequentiality in a little more detail. We can make the account recognizably realistic if we consider simple actions that we might observe in other animals. For example, if an animal with two forward facing eyes (organized for stereoscopic vision) is observed looking around (and the animal could of course be another human) we recognize that it does this simply because it cannot look everywhere at the same time. Typically, such an animal may look to one side and then another, and there is abitrariness in the selection of which direction is scrutinized first, but there is no avoiding the fact that one direction must indeed be first. There is a simple physiological constraint operating here – without completely different visual apparatus (for example, as found in many birds; see Birkhead, 2012), stereoscopic animal vision will operate sequentially. Indeed, it might be argued that the visual systems of some insects, which provide a much better approximation to omni-directional vision than the two eyes of a mammal, are the way they are precisely because they obviate the need for so much head turning. In contrast, the vision systems of mammals might be considered optimal for hunting, or alternatively, configured for searching for food whilst keeping “an eye out” for predators. Many birds manage well with two eyes which provide vision covering almost a 360° field of view (horizontally), but some have vision more akin to that found in mammals (e.g. owls). Stereoscopic binocular vision requires lots of head-turning for effective use (think of owls again; but see also Birkhead, 2012) and that means sequential activity linked to planning, following of prey, etc. It is worth noting here that the sequentiality is inevitable given the vision systems concerned; it is physiologically driven *inherent sequentiality*. The animal has no choice except in regards the selection of one arbitrary sequence instead of another. There may be social reasons or personal habits which determine the outcome (cf. the chant once taught to children in UK in relation to crossing a road: “look left, look right, look left again”) but these are not inherently part of the visual survey.

1(1).1.2 We turn now to a different sort of sequentiality, one where the sequence itself is specified: *inherent sequence*. Illustrations of this sort of sequencing in behaviour tend to involve the structure or physics of the world; they show the impact of the environment on an organism. Inherent sequences are properties of objects and the environment, typically. A squirrel cannot bury a nut without first excavating a little hole to push it into; the covering up and “making it all look undisturbed” comes last. There is an *inherent sequence* here, and it is unavoidable. Interestingly, when one observes the last phase of the operation it looks as though it is itself an illustration of *inherent sequentiality*. The squirrel opportunistically, as it were, finds little scraps of leaf debris, bits of grass (which it may have dug out in the first place), and so on, and carefully arranges it to look natural and undisturbed (and it does). Were the squirrel to operate at this moment with any sort of fixed sequence it would end up with something approximating the squirrel equivalent of a cairn. This would be characteristically artificial, and thus recognizable by other animals – a "squirrel was here" type of marker; not helpful.

1(1).1.3 We can be a bit more subtle in uncovering how *inherent sequence* fits with *inherent sequentiality*. Consider movement – specifically the use of limbs to crawl or walk about on land. It is self-evident that the use of legs for walking and crawling involves sequential, patterned, movement. The motion of each leg is itself the result of carefully sequenced and co-ordinated muscle contractions (an example of inherent sequence). The muscles in the leg do not all contract in one spasm (imagine a leg cramp, but much, much worse). Instead they work together, but sequentially, to produce the desired motion of the foot/ pad/ hoof removing it from the ground after a bit of a push (rather than just a lift) to maintain forward momentum, and then moving the limb to position the foot/ pad/ hoof for the next contact with the surface – thereby achieving a stride. The stride follows the inherent sequence, even though over rough ground the fine details are contingently governed by the terrain (this is use of context, as we will discuss later). All the while the other legs are being moved in co-ordinated sequentiality which can be called simultaneous in one sense (all the legs may be involved in locomotion) but sequential because each leg has its own movements aligned with the others so that the overall sequence is of the feet (pads, hooves) touching the ground in a managed pattern.

The variations in locomotion within an individual, within species, or from species to species are not relevant. Some quadrupeds walk with a diagonally paired gait, some with ipsilateral pairing, and some, indeed, can be trained to switch mid-stride. This scale of patterning may be hardwired and thus look a bit like inherent sequence, but equally may not. Either way it is certainly a reflection of inherent sequentiality. Likewise, do all the feet leave the ground at any moment? They might indeed in some species, but not others – it simply doesn't matter. The sequential imperative operates in both scenarios. Two feet, four, 8, 100? Again it doesn't matter – the requirement is that the operation is sequentially patterned. What seems clear is that inherent sequentiality is part of locomotion; inherent sequence is part of each stride (regardless of the number of legs).

1(1).1.4 Locomotion, like breathing and chewing, also structures the necessary sequentiality in cyclical form – so there is inherent cyclicity as a form of inherent sequence, and also sequentiality. This becomes relevant when we consider syllabicity in speech, thought to be exploitation of mandibular movement cycles in chewing. We return to this later, in both Part 2 and Part 3.

a tea-bag. The resources may or may not be coherently or tidily organized, and items may or may not be put away after use. There is arbitrary sequence in the process of assembly, and also arbitrary variation in how that sequence plays out to completion. Each tea-making event is worked out "on the fly" in a contingent manner. The boiling water – perhaps from a kettle, or from a dispenser in the kitchen – is poured into the mug and the resulting cup-full of proto-tea may be taken to a counter or table somewhere so that a spoon can be used to pound the tea-bag to extract as much flavour as possible – and to ensure the sugar is dissolved – before the tea-bag is removed and discarded (a fork handle would serve just as well if a spoon could not readily be located). The mug of tea is taken away to be drunk. Note that even here the inherent sequence involved in brewing the tea is preserved – that is fixed by the physics and chemistry of producing a brew.

1(1).2.2 The account above (somewhat laboured perhaps) of three completely different styles of making tea reveals essentially the same arbitrariness in the activity other than where it is constrained by inherent sequence. Each style is immediately comprehensible even if it is not *your* style.

1(1).2.2.1 One important idea to take away from this example is that the arbitrariness does not undermine the sense of purpose. The inherent sequence suffices as a thematic constant which supports the sense of purpose irrespective of the arbitrary details such as ordering of sub-components, retrieval of constituent items, and so forth. The coupling of arbitrariness with a sense of purpose surfaces again when we consider speech.

1(1).2.2.2 The second important idea concerns the details in the accounts. It is obvious that the scale of description fits the narrative purpose of differentiation – much less detail and the differences will be obscure. To be sure, they could be less detailed and still work to differentiate each style (typical kitchen tea-making process; pedantic tea-making ritual; student's tea-bag "dunk and go" cuppa) but then the reader would be left to supply/ infer much of the missing, differentiating, detail. On the other hand, offer even more information and the whole account will be submerged in details about movement of the hands, which hands are used to hold the spoon or the caddy, open a drawer or cupboard, and so forth. The descriptions are already tediously detailed, but could in fact be much more detailed. The importance of detail in providing accounts of behaviour should not be underestimated but the difficulty is picking the right scale of detail to illuminate the point being studied. We will find, in accounts of speech, that detail is really significant.

1(1).3 What we see from the foregoing is that sequentiality is unavoidable in behaviour. There can be mechanical ways around some of the sequencing, but nonetheless five points are clearly of general validity and interest: sequences can be inherent in an activity; sequentiality can be inherent; cyclicity can be found in both sequences and sequentiality (locomotion, eating); simultaneity is sometimes found woven into the sequentiality (and can be thought of as coordinated sequencing); arbitrariness is found where sequences are not specified. However, in all cases sequence can readily be exploited, ritualised and interpreted; a necessary sequence still requires planning, perception and contingent action and cannot be considered in isolation from the general case of managing sequential activity. Managing perception and the stripping out of time from sensory input has to be done regardless of the activity being observed or monitored. Inherent sequence seems special, and describing it as different from inherent sequentiality, and both as different from arbitrary sequencing, may seem to confuse the discussion. The point is just that they are all aspects of sequentiality which is unavoidable. The differences have to be learned and perhaps exploited, but underlyingly sequentiality is just that – the organism faces the unavoidability of the Sequential Imperative and the organism's brain has to deal with it.

1(1).3.1 What we have not discussed so far is the internal representation of the activities to be performed. This will unfold as we progress, but here it is helpful to point out that in a general sense the representation of an activity (visual search, locomotion, whatever) is not itself an activity. The representations within the brain are, in essence, atemporal.[2] The functional specification of the brain is therefore concerned with serving the Sequential Imperative by managing the transfer of representations into (and out of) the temporal domain. My desire for (or plan for the making of) a cup of tea does not itself unfold in time, but it is expressed sequentially. We will encounter this "time transformation" notion again (it is the *curious sort of time machine* noted in the Introduction, but see also 1(2).5.2, 3(7).13.4, and various other points). Here it is enough that the reader recognizes the core idea. Notice that where the sequences in behaviour are arbitrary, it is easy to see how meaning can be attributed to the sequences deployed. Further, the deployment of sequentially patterned activity solely for the purpose of providing for the attribution of meaning (bird song, human speech) is merely an extension of this attribution of meaning to activity. Although there is much more detail to consider (and see Part 3) we are already relating general aspects of behaviour to language.

2 Later on we will see the need for a more nuanced account (see also 1(1).4.7.2).

generality is simple, the detail, as ever, more challenging.[5] Notice, in passing, that we have not so far mentioned the mind. The functional specification of the brain is about the brain – any brain – not the human mind. We will return to the notion of mind in the third Part of the book.

1(1).4.9 One point needs to be noted for consideration for later – and those familiar with debates in AI and Cognitive Science, as well as speech processing, will have spotted the point. Segmentation was briefly mentioned above and was characterized as imposed on the continuity of action. The mention was necessary there to quell unease that an assumption is being made about the nature of segments or their ubiquity. The issues are non-trivial and will be followed up in the mid-Part of the book (2(5).10.1.5).

1(1).4.9.1 Lastly, on this account homeostasis need not have been a prior aspect or focus of the functionality of the brain but may in fact have been so. In any case it can be considered to be the *activity* (patterned in time) which maintains an organism's state, or manages a transition from one state to another state, contingently in response to some internal representations of states and the environment. Homeostasis is governed by the Sequential Imperative as much as other activity and is thus as concerned with transformations into and out of the temporal domain as is making a cup of tea or describing a house. But as noted earlier – this book is not about homeostasis. One implied project coming from this work could well be to extend the account to cover homeostasis in an evolutionary context.[6]

---

5 It could be argued that the notion of cybernetics, which was formulated in 1948, came to be thought of as a functional specification later, but work in this vein has yielded very little (see Clark, 2013, and the discussion in 2(6).11.6). For a review of Cognitive Science, its component disciplines, and its history, see *The Cambridge Handbook of Cognitive Science*, (especially Pt 1). Frankish and Ramsey (2012).

6 See Damasio and Carvalho (2013); also 1(3).7.9.1 and 3(7).13.6.3.

1(1).3 What we see from the foregoing is that sequentiality is unavoidable in behaviour. There can be mechanical ways around some of the sequencing, but nonetheless five points are clearly of general validity and interest: sequences can be inherent in an activity; sequentiality can be inherent; cyclicity can be found in both sequences and sequentiality (locomotion, eating); simultaneity is sometimes found woven into the sequentiality (and can be thought of as coordinated sequencing); arbitrariness is found where sequences are not specified. However, in all cases sequence can readily be exploited, ritualised and interpreted; a necessary sequence still requires planning, perception and contingent action and cannot be considered in isolation from the general case of managing sequential activity. Managing perception and the stripping out of time from sensory input has to be done regardless of the activity being observed or monitored. Inherent sequence seems special, and describing it as different from inherent sequentiality, and both as different from arbitrary sequencing, may seem to confuse the discussion. The point is just that they are all aspects of sequentiality which is unavoidable. The differences have to be learned and perhaps exploited, but underlyingly sequentiality is just that – the organism faces the unavoidability of the Sequential Imperative and the organism's brain has to deal with it.

1(1).3.1 What we have not discussed so far is the internal representation of the activities to be performed. This will unfold as we progress, but here it is helpful to point out that in a general sense the representation of an activity (visual search, locomotion, whatever) is not itself an activity. The representations within the brain are, in essence, atemporal.[2] The functional specification of the brain is therefore concerned with serving the Sequential Imperative by managing the transfer of representations into (and out of) the temporal domain. My desire for (or plan for the making of) a cup of tea does not itself unfold in time, but it is expressed sequentially. We will encounter this "time transformation" notion again (it is the *curious sort of time machine* noted in the Introduction, but see also 1(2).5.2, 3(7).13.4, and various other points). Here it is enough that the reader recognizes the core idea. Notice that where the sequences in behaviour are arbitrary, it is easy to see how meaning can be attributed to the sequences deployed. Further, the deployment of sequentially patterned activity solely for the purpose of providing for the attribution of meaning (bird song, human speech) is merely an extension of this attribution of meaning to activity. Although there is much more detail to consider (and see Part 3) we are already relating general aspects of behaviour to language.

2 Later on we will see the need for a more nuanced account (see also 1(1).4.7.2).

1(1).4 Before moving on to consider how all this can be captured systematically and in a way which leads to further insights, it is worth noting that some of the concerns elaborated above can be identified in the conventional Cognitive Science literature, including work relating to communication.

1(1).4.1 Neisser (1967:140), discusses the integration of saccades (the rapid sequence of brief visual fixations which are a normal part of vision) – the "snapshots" in the extract below – in visual perception. He describes what we will be calling de-sequencing but does not take the extra step of pointing out that the sequencing and de-sequencing of behaviour and perception are *necessary* because the organism has to conform to the Sequential Imperative. It is almost as if he has half got hold of the core idea, but doesn't see the imperative.

He writes:

> ... if we see moving objects as unified things, it must be because perception results from an integrative process over time. The same process is surely responsible for the construction of visual objects from the successive "snapshots" taken by the moving eye.
>
> Under normal conditions, then, visual perception itself is a constructive act. The perceiver "makes" stable objects, using information from a number of "snapshots" together. Such a process requires a kind of memory, but not one which preserves pictorial copies of earlier patterns. Instead, there is a constantly developing schematic model, to which each new fixation adds new information. The individual "snapshots" are remembered only in the way that the words of a sentence are remembered when you can recollect nothing but its meaning: they have contributed to something which endures. Every successive glance helps to flesh out the skeleton which the first already begins to establish.

1(1).4.2 Another illustration of the Sequential Imperative and what is entailed when serving it is readily brought to mind if one considers how one would go about describing the arrangement of rooms in one's home (cf. Levelt, 1993:139ff where he discusses Linde and Labov, 1975). One has a "mental image" of some sort, or representation of the static arrangement of rooms, and that representation is itself atemporal. Although the rooms don't move about, one has to mention them in one or another sequence – and so the question is what sequence do you use? Some sort of imagined tour might be contemplated, but the explanation does need to be anchored so that the tour does not flit miraculously between floors, for example. Alternatively, a ground plan might be described, along with the relative location of rooms on the plan, floor by floor (and thus does permit some flitting between floors). The speaker has to linearize their internal representation of their home in a way which eases for the listener the problem of creating (de-linearizing) an internal representation or image of the home. There is arbitrariness in the language and in the selection of an order for delivery of the account, but that delivery is a sequential affair. This illustrates the time transformation notion referred to above.

1(1).4.3 Levelt's (1993:138ff) discussion of the above experience is set out as a facet of what he terms "the linearization problem" which is the speaker's need to decide the order in which topics are mentioned in discourse. Typically the listener knows that because the order can be significant in one way or another, and also that the speaker knows about this, s/he can expect the speaker to use order to achieve effects or specificity in meaning. This is rather more general than picking the order in which to describe rooms in a house. In one form the problem can be concerned with the pragmatics of the situation – topicalization may lead to a particular ordering which doesn't change meaning so much as emphasis.

Levelt's example is the following two sentences:

I will send you the money next week.
The money I will send you next week.

The meaning doesn't differ – the pragmatic force is different. This can be contrasted with adherence or not to natural or implied order:

She married and became pregnant.
She became pregnant and married.

These two sentences have quite different meanings because it is assumed that the speaker and listener understand that the two topics have a temporal ordering matching their order of mention (and the difference can be socially interpreted).

1(1).4.4 Just as Neisser seems to grasp part of the sequential imperative, as expressed in his account of stripping sequence out of a sentence to leave just the meaning, so it appears that Levelt has grasped a different part. Levelt's concern is about the pragmatics of language use, and how to organize sentences or parts of sentences into sequences which match intentions. This is putting the sequence into the language, rather than stripping it out. Both "directions" of working with sequence are important, as we will see. Furthermore, the issues actually go beyond language use to cover all behaviour in all animals with brains. This, then, is the topic of this book: there is a Sequential Imperative and the brain is the organ which serves it, or "deals with it", or manages it – the core idea is simply stated but the details turn out to be very complex.

1(1).4.5 The material of this chapter should have caused the reader to recognize that organisms necessarily organize both their activity and their perception sequentially, irrespective of the presence or otherwise of sequential specification (inherent sequence) or sequential pattern (inherent sequentiality and cyclicity). It is not enough just to assume that stuff happens – somehow – and that the real task of Cognitive Science is to work out complex problems of vision, linguistic meaning, philosophy or semiotics. In the next chapter we take a more formal view of how the Sequential Imperative places functional requirements on creatures and their brains.

1(1).4.6 We have discussed activity in rather general terms, but with an implication that it is segmented on the scale of interest (scale of relevant description, as in the account of tea-making). In fact activity is continuous, even when it seems segmented. Continuous movement, as in gliding flight, swimming underwater, or snake slithering, involves continuous un-segmented activity even if oscillatory. Sequencing and de-sequencing are conveniently discussed in terms of segmented activity – as we will see – but segmentation is not required for the notion to make sense. In terms of complex behaviour segmentation does indeed make sense, but even in speech there are aspects of what goes on which reflect continuity of action, whether in the case of transitions between segments, or control of timing (see 2(4).8.1.2.5). It is perhaps convenient to think of segmentation as imposed on continuity (see also Chapter 5).

1(1).4.7 In summary thus far – the Big Idea is that the brain serves the Sequential Imperative (SI), in the sense that the heart is for pumping blood. The brain doesn't (need to) do anything else.[3] What we have seen so far is that the sequencing and de-sequencing link the "internal world" to the "external world" and the brain manages the transformation into and out of the temporal domain. In the next chapter we will formalize this in the shape of General Cognitive Principles (GCPs). GCPs look like detailed note-forms for specific issues to be addressed by Cognitive Science. In later chapters we will look in more detail at how exactly the brain must serve the SI – there are many details to be worked through to make the core idea viable or coherent – but thus far the

3 One part of the brain, the hypothalamus, is involved in homeostasis through the autonomic nervous system and the endocrine system. It is conceivable that an evolutionary account can be provided which shows that even such "basic" brain activity nonetheless reflects the general functionality of the brain. See also, for example, 1(1).4.9.1.

crude picture is enough; the brain transforms intentions, beliefs, desires, plans, etc. into sequenced activity, whilst at the same time, and in coupled processes, it takes a constantly varying sensory input and transforms this out of the temporal domain to yield percepts, images, etc. The processes of what we will call *sequencing* and *de-sequencing* are complex and require careful management, as we will see in Chapter 7.

1(1).4.7.1 It is important that the reader "gets" the idea of the Sequential Imperative and its significance for the brain and for any organism that has a brain. If a creature is to behave and perceive then it must manage the bi-directional time transformation between the sequentiality in the external world, and the atemporal internal world of perceptions, ideas, plans, memories, and all the rest. That is the Sequential Imperative. That's what the brain is for, and indeed that is all that the brain is for (this may seem a bit bleak; we will return to his point).

1(1).4.7.2 The Sequential Imperative is characterised thus far as being the necessary bi-directional transformation between "on the outside" unavoidably temporally patterned activity and perception, and "on the inside" the atemporal plans, intentions, memories and etc., which are stored in the brain. There is, of course, a sleight of hand here – some of the "in the head" cognitive entities may be time varying (as we know when we learn new facts or activities, and as we sense when we deliberate). But crucially, this time varying aspect of the "in the head" entities is not coupled directly to the external activity and perception. There is still a bi-directional time machine, but it is now a little more subtle. The simple model will suffice until much later in the book when we look in detail at learning and thinking.

1(1).4.8 Nothing has been said yet about the size of brain or being.[4] If there is a brain, however small, it must serve the Sequential Imperative. This is the big picture – the Cognitive Science being worked on is pan-species. I think this is really bold as a claim, but also very simple.

It would seem that no-one heretofore thought to tackle the question: "What is the functional specification of the brain?". We are doing that now – the

4 Here, as elsewhere in the book, brain size is used loosely to indicate complexity. The relationship between brain size and complexity of neuronal arrangement and density remains an open question. Recent work shows something of the complexity; see for example the review by Steven Mithen of *The Human Advantage: A New Understanding of How Our Brain Became Remarkable* by Suzana Herculano-Houzel. MIT Press, 2016. *The New York Review of Books*, **63**(18), November 24, 2016.

generality is simple, the detail, as ever, more challenging.[5] Notice, in passing, that we have not so far mentioned the mind. The functional specification of the brain is about the brain – any brain – not the human mind. We will return to the notion of mind in the third Part of the book.

1(1).4.9 One point needs to be noted for consideration for later – and those familiar with debates in AI and Cognitive Science, as well as speech processing, will have spotted the point. Segmentation was briefly mentioned above and was characterized as imposed on the continuity of action. The mention was necessary there to quell unease that an assumption is being made about the nature of segments or their ubiquity. The issues are non-trivial and will be followed up in the mid-Part of the book (2(5).10.1.5).

1(1).4.9.1 Lastly, on this account homeostasis need not have been a prior aspect or focus of the functionality of the brain but may in fact have been so. In any case it can be considered to be the *activity* (patterned in time) which maintains an organism's state, or manages a transition from one state to another state, contingently in response to some internal representations of states and the environment. Homeostasis is governed by the Sequential Imperative as much as other activity and is thus as concerned with transformations into and out of the temporal domain as is making a cup of tea or describing a house. But as noted earlier – this book is not about homeostasis. One implied project coming from this work could well be to extend the account to cover homeostasis in an evolutionary context.[6]

5 It could be argued that the notion of cybernetics, which was formulated in 1948, came to be thought of as a functional specification later, but work in this vein has yielded very little (see Clark, 2013, and the discussion in 2(6).11.6). For a review of Cognitive Science, its component disciplines, and its history, see *The Cambridge Handbook of Cognitive Science*, (especially Pt 1). Frankish and Ramsey (2012).

6 See Damasio and Carvalho (2013); also 1(3).7.9.1 and 3(7).13.6.3.

CHAPTER 2

The Sequential Imperative – II

1(2).5 In this chapter we look at a more formal account of the Sequential Imperative (SI) – captured in some General Cognitive Principles (GCPs). The first three principles set out here (with more to come in the next chapter) capture the essence of the Sequential Imperative. All the principles are provided here *ex caeruleo*, without derivation or deduction – as if they have sprung fully formed onto the page. Explanation is offered in relation to the SI, and thus justification. The principles should seem obvious, or at least blindingly plausible, by the time the reader reaches the end of this chapter.

We also consider later in this chapter some early neurophysiological work which might be thought relevant – Lashley's influential paper on cerebral mechanisms and serialization (Lashley, 1951) – and work which followed in the next 10–20 years, showing that many of the component ideas and data were in place for understanding the SI, but the conceptual drift went elsewhere. In some senses, therefore, the ideas presented now could have been developed some time back – we are not dependent here on recent radical thinking, or new data from esoteric instrumentation (see also Frankish & Ramsey, 2012). The relevance of this earlier work is simply that it addresses the obvious question – is it really the case that goals, intentions, plans, memories, percepts, etc., are atemporal representations in the brain? The question matters, of course, because as noted earlier in relation to time-switches the details of the interior mechanism must not be relevant to appreciation of the functionality. *We can then work out what the brain does without knowing how it does it.*

The first three principles are as follows:

GCP1 Sequentiality in behaviour is forced physiologically.
  Corollary 1: Sequence penetrates the corporeal boundary.
  Corollary 2: Sequence is semiotically free.

GCP2 Cognitive entities are:
  i) Inherently atemporal
  ii) Dual in nature

GCP3 Behaviour is sequencing; Perception is de-sequencing.

1(2).5.1 We begin our more formal look at the Sequential Imperative with GCP1. This has three components, the first of which simply captures the previously described necessity for sequence – see for example 1(1).2/3 above. This may

© KONINKLIJKE BRILL NV, LEIDEN, 2017 | DOI 10.1163/9789004342996_005

seem to be just a summary, and thus somehow "enough". That the SI has to be elaborated further is our focus in this chapter. Indeed, in subsequent chapters we will work through some of the ramifications of what here is stated rather simply. Notice also that the type or style of sequence (as elaborated in the previous chapter to illustrate the ubiquity of sequencing and de-sequencing) whilst clearly relevant to understanding what an organism is doing at any moment is not actually relevant in terms of the Sequential Imperative. The brain has to manage the SI regardless of the style of sequence involved.

1(2).5.1.1 Sequentiality is more subtle than might at first appear. We are considering strictly the "A before B before C…" – ness of activity and sensory stimuli. GCP1;Cor1 states that sequence "penetrates the corporeal boundary". What this means is that a sequence of muscle firings, and consequent actions "in the external world" map one to one. To be sure, there will be some timing differences, and clusters of muscles activated more or less simultaneously, but the overall effect is – indeed must be – that the activity maps sequentially onto the "instructions" internally which produce the activity. Likewise, sensory feedback about the actions will match the sequence of those actions and thus the sequence of the "muscle instructions" to perform the actions. Monitoring of motor activity and performance would become impossible if the sequential mapping did not exist.[1] In essence – a sequence externally, but distant (distal) is also a sequence externally at the sensors (proximal). The sequence within (proximal – the immediate output of the sensors) is the same; the time transformation mentioned earlier (1(1).3.1) relates the interior proximal to the interior distal. The corporeal boundary separates the external proximal and the internal proximal, and it is the sequentiality which penetrates that boundary (photons or soundwaves, for example, do not somehow pass directly into the brain and soundwaves do not emanate from the brain directly; the production of speech sounds is detailed in Chapters 4 & 5).

1(2).5.1.2 GCP1;Cor2 states something else entirely – but no less important. The fact that sequence is forced means that it is inherently meaningless, and thus free to be given meaning – sequence is free semiotically (this is the case whether or not the sequentiality is inherent sequence or just sequentiality). This provides for the possibility of language whilst not obliging all organisms to have

1 The quotation mark induced sense of vagueness need not perturb the reader. The cautions serve to remind one that although there is detail to be provided (Chapters 4 and 5 especially) its absence here doesn't break the argument.

language. Just as we can recognize meaning in the manner in which someone makes tea, so, eventually, we can understand how patterns of activity which are essentially arbitrary (the sounds made by the human vocal tract, say) can be strung together in short arbitrary sequences (morphemes) co-ordinated in yet other arbitrarily structured sequential patterns (syntax) to make what we call speech. (See Part 2.) Note that in all cases the sequentiality, coupled with Cor1 and Cor2, permits continuous re-internalisation of the ongoing externalisation – monitoring and control of the activity so that it matches intentions. This crucial bi-directionality is operative at all scales of activity – and this can be recognized with anecdotal examples. Consider a sculptor using a chisel of some sort on a block of wood. The activity is constantly managed and monitored (sequenced and de-sequenced) at the scale of individual acts of removal of fragments of wood; at the scale of fashioning a specific part of the intended object – say the eyes in the expression on the face; at the scale of the proportion of the whole head to the rest of the body (in an effort to control the exaggeration of the eyes in the expression); at the scale of the head in relation to the length of the body and the sweep of the tail; and then there is the bone balanced on the dog's nose; and so on and so forth.

1(2).5.2 We turn next to GCP2. This is a little less obvious than GCP1 but actually quite straightforward.

i) The core concept is that ideas, beliefs, desires, intentions, memories… are brain states, not time-varying patterns of neural activity aligned with or entrained by the "external world". Our awareness of these cognitive entities may have temporal characteristics, but those are not their inherent nature (and of course in some creatures the notion of awareness is vague or even absent but the brain states remain just that – we return to this in Part 3). My desire for a boiled egg does not simmer in my head for 4.5 minutes (approximately). The desire is about something with a 4.5 minute duration (the cooking of the egg) but does not itself have such temporal structure – it merely endures (cf. the quotation from Neisser in 1(1).4.1 above). Thus we can say that cognitive entities are atemporal. A rather more nuanced account is developed later, but for now it is enough to consider that the entities are not time varying in any sense related to the sequentiality of activity or perception. We also need to keep in mind that the account being developed is not about just humans – so boiling some eggs, or having a belief in (a) god, is not definitional or necessary. In fact, as we will see later (e.g. 1(3).7.6), although it seems readily appreciated that a desire is not sequentially organized, the atemporality operates at all scales. Recognition of a familiar face, memory of a bird-song, or knowledge of the meaning of a particular word, and in fact all other cognitive entities (in all species) do not imply or require a temporal unfolding in the head; they endure even if about a pattern in time.

ii) However, for sequencing to be possible the internal representation of the cognitive entity for a desire (etc.) has to have or specify component cognitive entities each of which has components.... not quite "all the way down" but necessarily to the point at which physiological instruction, delivery and monitoring can be divided no more finely (the proximal interior). This componentiality[2] is more insightfully captured as *duality* whereby any cognitive entity can be envisaged as not only having components but also being a component. These components, when sequenced in behaviour, are the segments noted earlier (1(1).4.6).

1(2).5.3 The third GCP now emerges as unifying of the insights just described. Behaviour, in accord with the Sequential Imperative, requires the unpacking of components, at ever finer scales, to provide a sequencible output expressive of a memory, desire, belief, etc. Likewise, perception requires the reverse process – de-sequencing – to build up bigger entities from smaller ones, to yield eventually a memory or internal image, or whatever, which is a state with duration, not a sequence (reflect again on Neisser's description of this process, in the quotation offered above (1(1).4.1)).

1(2).5.3.1 GCP3 requires some illustration to make its significance clear. For this we can draw on something from human behaviour – language – but the principle applies to non-human brains (as do all the principles). Typically one encounters discussions of the structure of activity, or of the product of activity, as being hierarchically organised. So one finds that speaking a sentence involves sub-component entities such as phrases, composed of words, and so forth, in a hierarchically arranged structure. These are frequently referred to as "tree diagrams", with the root at the top.

2 I attempt in this book to control the impulse to create neologisms – these can arguably be considered a sign of a writer's poor vocabulary/ skill. It is justified here. *Componentiality* refers to duality, as just explained – being a subcomponent and having subcomponents. *Componentialities* refers to the fact that an entity – say a word – can have sub-components in several different domains (phonetic segments, syllables, morphemes, stress bearing elements) which do not map onto each other; likewise it can be a sub-component in several structures (syntax, stress). This captures the sense that for an entity the links to components (in both directions) are at least multiple and may also be different from time to time, and context to context.

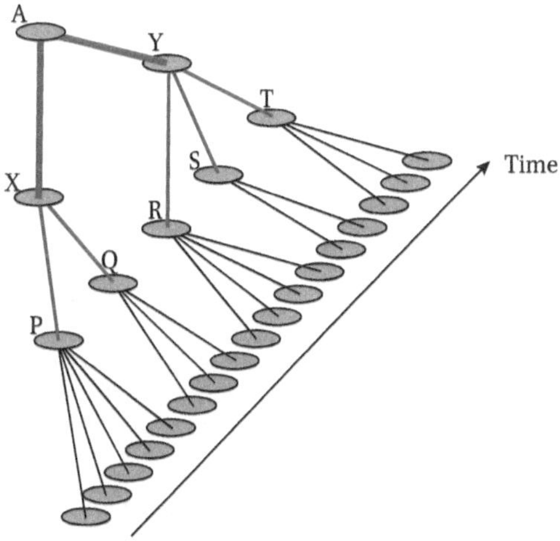

FIGURE 2.1 *Conventional tree diagram showing the surface elements (sounds) aligned with the flow of time and linked to the nodes in the next level up in the diagram (P-T) and then those nodes linked to X&Y and thus to the root A.*

1(2).5.3.1.1 Consider a sentence: *Stan's dog barked at him.* Let us ignore for this exercise the detail implied by the possessive marker on *Stan* (some of this is explored later, in Part 2). In the diagram (Figure 2.1) it is the sounds which surface as the sequentially arranged elements.

The sentence can simply be diagrammed in a conventional way but note that although the size of the elements in the temporal stream is shown as uniform this is merely a graphical convenience – no assumption is made of uniformity of segment duration in the speech stream. The root, or node A, is typically labelled S for sentence, X and Y become Noun Phrase (NP) and Verb Phrase (VP) respectively, the five words map to P -> T, and the sounds of the words map onto the little blobs on the timeline. The sounds of the words are sequenced, that does not seem in doubt (duration of each sound is a separate issue, addressed later). What about the words? The diagram shows them in sequence, as one might expect from the sounds. Likewise the NP and VP (X and Y) are in sequence. Much emphasis is placed by linguists on the style of these diagrams, where the *sequential* arrangement of the nodes is set out, and lines showing connections between components on two levels may not cross – so Y could not connect to Q, with X to R, for example.

1(2).5.3.1.2 Consider next the figure below (Figure 2.2) which shows how it is possible to illustrate the same structural relations as in the illustration of the conventional hierarchy, above, but with the assumption that the time dimension (sequentiality) is reflected *only* in the output sequence of sounds. Here we see

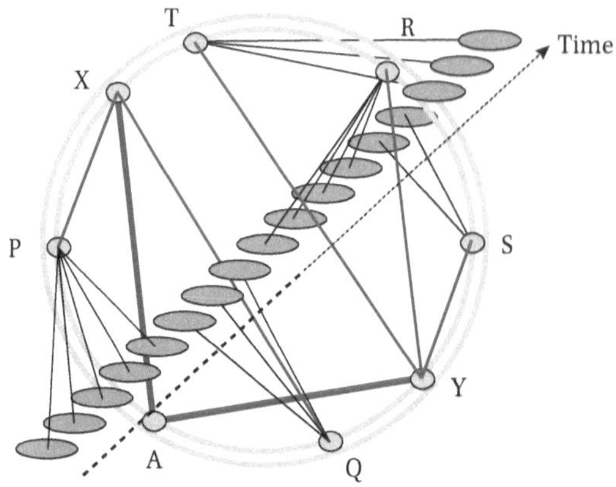

FIGURE 2.2 *Unconventional diagram showing structural nodes arranged in a circular fashion with links. The surface elements are arranged in a linear fashion aligned with the flow of time (as in Figure 2.1).*

that the nodes A, X, Y, P, Q, R, S, T are all arranged in a circular fashion around the sequence of output sounds – these nodes are simultaneously co-present precursors or specifications of the output. The diagram resembles a Naum Gabo sculpture with links (strings) between structures in different "planes".

The different "planes" are defined by the central "spine" of sequenced output segments running along the time axis, and the sets of links to the various nodes P, Q, R, S, T, so that the links are in different planes radiating out from the central spine in different directions. An alternative visual aid for these structures is to think of a notebook held open with all the pages fanned out around the spine. The notebook has a "spiral" binding, or set of rings, where each ring in the spine of the book represents a speech segment, with pages, or planes (half-planes, strictly) fanned out away from the spine. Specifications written on the pages (P, Q, R, S, T) label the linked segments in sequence but are themselves unsequenced (the pages fanned out around the binding are not sequenced, but contain specifications about the segment sequences in the output).

1(2).5.3.1.3 The crucial point to understand in the second illustration above is that the intermediate scale nodes P, Q, R, S, T, X, Y, along with the root node A, are all co-present precursors of the eventual output sequence. Note that for diagrammatic simplicity node X is not shown linked to the first 8 surface sequence components, which in reality it is. Those surface nodes have quality X as much as quality P or Q; they are all part of the surface delivery of the subject noun phrase. Likewise, Y's links to the remaining 9 surface sequence components are not shown – their "Y-ness" is significant, but clutters up the diagrams. Note also the use of the word *scale* in these discussions, rather than the more conventional word *level*, which is perhaps better suited to the more conventional tree diagrams.

1(2).5.3.1.4 Another important point is that the two diagrams offer different perspectives on the data – the utterance. The more conventional diagram can be read as an account of the formal structure of the sequence of sounds, although more than one reading is possible, and complicated claims can be made which seem hard to support (e.g. the inadmissibility of structures with crossing lines – noted earlier – is a statement about the linguistic model being deployed rather than about the cognitive processes of producing a linguistic utterance). The second diagram is more acceptable as representational of what is going on in the head of the speaker. The precursors of speech are assembled in coordinated co-occurring interrelated specifications of different scales of structural detail. However, the diagram is difficult to read. This issue is taken up in Part 2.

1(2).5.4 Thus far the discussion of the General Cognitive Principles has focussed on the way they reflect or encapsulate the Sequential Imperative. A model is provided to illustrate an additional point, one that is entailed by the core idea of the brain serving the SI, namely that the internal atemporal entities are brought to bear on the output stream (and indeed the input stream, but that is not illustrated) in a co-occurring and non-hierarchical fashion. This will be the focus of the material of Chapter 5.[3] However, the reader will need reassurance about the claim that internal representations are atemporal – how can we make sense of this claim? We deal with this, and other neurophysiological issues, next.

1(2).6 We now turn to an issue raised earlier (1(2).5.2), the neurophysiological factors which need to be considered in connection with GCP2 & 3. These can usefully be considered in three groups – those which arise from Lashley's discussion of "Serial Order in Behavior" (Lashley, 1951); those which arise from consideration of neurophysiological temporal factors, discussed primarily in the 1960s; and those which arise from consideration of the need to match fine scale specification of action to actual muscle contraction (assuming it is not

3 Part 3 will pick up an issue not addressed here – namely the difference between hearing and vision when discussed in terms of GCP3. There is modality dependent asymmetry between perception and behaviour when considered in terms of sequencing and desequencing (see for example 3(7).13.4.4.3).

inherently and invariably specified anatomically) – this amounts to the problem of calibrating the link between intention and effect.

1(2).6.1 Lashley's concern (1951) is to show that conventional behaviourist models cannot account for serial organization in behaviour.[4] He pointed out that a stimulus "is never into a quiescent or static system, but always into a system which is already actively excited and organized". And further, aspects of both stimulus and pre-existing activity in the system will be at many different scales (in time). Language behaviour demonstrates this very clearly, as he pointed out. He noted that temporal integration (reconciling and coordinating the activities and stimuli on different scales) "is not found exclusively in language; the coordination of leg movements in insects, the song of birds, the control of trotting and pacing in a gaited horse, the rat running the maze, the architect designing a house, the carpenter sawing a board present a problem of sequences of action which cannot be explained in terms of successions of external stimuli".

1(2).6.1.1 Lashley goes further than simply pointing out that attributing to the cerebral mechanisms some sort of content free capability, linking stimuli to responses, and responses to stimuli, can't work because the brain is busy with many tasks and activities, all with different scales, all the time. His point is that the brain cannot be construed as serving or supporting a stimulus-response conception of serially ordered behaviour. The problem of serial order in behaviour is thus essentially a problem for behaviourism, as a model, once the details of neurophysiology become clearer. However, providing a neurophysiologically supported account of how the brain does deal with serial order in behaviour remained an issue in 1948/51. The atemporal nature of representations, and their transformation into sequences, fit well with notions of hierarchical structures (but see the discussion above (1(5).3.1.2)). Chomsky's independent development of this point encouraged people to see hierarchical organization in representations as a way to resolve the problem of serial order (he was not aware of Lashley's work, although both were at Harvard at the time, when

4 In September 1948 at the California Institute of Technology a symposium was held on the *Cerebral Mechanisms of Behavior*, under the auspices of the Hixon Fund Committee. This meeting – "The Hixon Symposium" – resulted in a book of papers, published in 1951 and re-issued in 1967. The contributing authors were W.C. Halstead, H. Klüver, W. Köhler, K.S. Lashley, W.S. McCulloch, and J. von Neumann; much of the discussion was also edited into commentaries on the papers. Lashley's paper is perhaps the most cited – it is titled: *The Problem of Serial Order in Behavior.*

he developed his ideas about linguistic structures (see Chomsky and Otero, 2004:95)). However, as illustrated above, hierarchies are not the only way to couple atemporal representations, which are dual in nature (having cognitive entities as components, and being components of other cognitive entities), to serial organization in behaviour. Indeed, in one sense a hierarchical structure is an imposed reading of a (simplified and incomplete) version of the output sequence specification captured in diagrams like Figure 2.2. The contrast is shown in the comparison of Figure 2.1 (a reading, or interpretation) with Figure 2.2 (the internal specification).

1(2).6.2 Since Lashley's time much work has been done on electrical activity in the brain and by the time Lenneberg wrote *Biological Foundations of Language*, in 1967, a considerable amount was known – enough for it to have been possible to follow up Lashley's concerns without being overwhelmed by concerns about language. In terms of the account being offered here, the relevance of these earlier findings is that temporal structuring of neural activity is complex enough for virtually none of it to be obviously linked to temporal patterns in stimuli or actions. The notion that somehow the internal representations of beliefs, goals, intentions, memories etc., might also be structured in time seems inconceivable, and so it makes sense to say that they are atemporal. It is an assumption, but it works.[5] We can consider some of the earlier work providing details of the various scales of temporal activity to be found in neural mechanisms. This is offered as illustrative support for the notion that the workings of the wetware do not impinge on the functionality of the brain, and that this was knowable in Lenneberg's time.

1(2).6.2.1 The neural mechanisms – along with their chemical environment – clearly exhibit temporal structures. It is possible to appreciate the temporal range of these without much difficulty, and often from quite gross evidence (in the sense established when the neuron is the working unit). Even though much of this evidence comes from the study of perception (much of it mid-20th Century work) it makes a valid contribution to our discussion.

5 For completeness I should also add that I am not considering the remote possibility that the general sense of electrical noise one encounters in neuroelectrophysiological measurements is the result of superposition of temporally structured electrical activity arising from internal representations of *everything* a person might know or believe. Such knowledge/ belief must surely be cast in neurological states, not activity, else the brains of humans would not show unlimited capacity for knowledge.

1(2).6.2.1.1 A look at temporal resolution in human binaural spatial location of a sound source shows that small time differences in the sound patterns reaching the two ears are interpreted as locational differences. I once contrived an experimental set-up in which two video-tape-recorders were synchronized in play-back to within two line periods (the UK TV standard at that time was 64 micro-seconds for each line period) with a small tracking error (around ±50 micro-seconds) which varied over a few seconds. Such precision was more than adequate for the visual purpose of having two synchronised views of an event. However, the two sound tracks of that event, when relayed unmixed to the two ears, produced a remarkable sensation of the sound source swinging from side to side in relation to a point which itself was offset to one side. Subsequent reference to a standard undergraduate psychology text revealed that binaural time differences as short as 30 micro-seconds are discriminable. The implication of this sort of observation, for our purposes here, is that neuronal activity can be structured on very short time scales.

1(2).6.2.1.2 Evidence for temporal patterning of sensory responses to stimulation comes from studies of hearing (Geldard, 1953). The patterns of discharge in the auditory nerve (of a cat) show that firing can be entrained by the stimulus up to around 850Hz. Between 850Hz and 1700Hz the evidence suggests that each fibre discharges only every second cycle. Above 3000Hz the synchronization is lost entirely.

1(2).6.2.1.3 Neuronal speed of response within the brain is in general no faster and may be slower (we are here distinguishing between the possibility of some special-purpose system for binaural timing differences and hearing, and more general-purpose neuronal activity). For example Rumelhart and McClelland (1986; Vol1: p130) cite a figure for the speed of operation of neurons as being measured in "milliseconds – perhaps 10s of milliseconds" – their purpose is to make a comparison with electronic computational hardware. Later (McClelland and Rumelhart 1986; Vol2: p366) we find:

> The average discharge rate of neocortical neurons is relatively slow – perhaps only 50 to 100 spikes per second, though the rate for a brief time can rise as high as several hundred spikes per second.

1(2).6.2.1.4 Other data concerning neuronal activity come from EEG records and from Evoked Potential studies. Both these latter techniques record signals extra-cranially, as voltages (electric potentials) between pairs of electrodes placed on the scalp. These techniques produce gross summations of the activity of many neurons. In the case of EEG these may be recorded as waves – so-called "brain-waves" – with frequencies of 1,3,5,7,10 & 13 Hz. (Lenneberg, 1967). In the case of Evoked Potential records the signals are recorded in response to a stimulus – perhaps a click – and averaged over many stimulus-response events. When this is done a clear electrical response to the stimulus is revealed (not usually discernible in the "noise" in an individual trace). Magnetic field equivalents of these potentials are also recordable (MEG, EF) and EF records, for example, can provide more precise indications (than EP) of the source of the signal in response to clicks.

1(2).6.2.1.5 Information about the source of the signal within the brain is also available from other techniques for imaging the brain. Tomographical techniques dependent on differences in blood-flow rates in various parts of the brain (perhaps using radioactive substances injected into the blood-stream elsewhere, but more recently using fMRI technology) can provide localization information, for example revealing sustained metabolic responses to stimulation over a period of seconds or minutes. (See also Lieberman, 2006; Chapter 4). Instrumentation is constantly improving, providing more resolution in space, time and connectivity. We know a lot more about what goes on in the brain but the question here is does all this work (and even computational modelling) matter? My answer is that it is not relevant to working out the details of the functional specification of the brain.

1(2).6.2.2 There are two points to note in connection with these various measures of neurological activity. The first is that activity is measurable over periods varying from a few milliseconds – or perhaps less if we recognize that binaural localization must require neuronal temporal sensitivities of a few tens of micro-seconds – up to periodicities of a second or so (slow "brain waves"), and to sustained levels which aggregate responses in specific cortical areas – e.g. in response to visual stimuli or to auditory stimuli. The second point is that in a few cases it is possible to link observed activity to the stimuli in a temporal fashion. With auditory stimuli this is clear – whether in terms of evoked responses or in terms of neuronal discharge entrainment or binaural time differences. More obscurely, perhaps, Lenneberg (1967:115–117) argues that gross measures such as the 7Hz brain wave are linked in a general sense to the capacity for finely controlled rapid action such as in speech, especially syllables.

1(2).6.2.3 The significance of all these observations and comments is that apart, perhaps, from Lenneberg's conjecture concerning 7Hz brain waves, it makes no sense to seek to align measures of temporal activity and patterns measured in the brain with temporal structures in behaviour and perception. This is in effect the detail that Lashley couldn't provide in his account of the inherent problem with behaviourism. In respect of Lenneberg's conjecture about the 7Hz brain waves Lashley might well have responded that such waves could deal with speech rhythm, but not its content. In respect of our earlier discussion of time-switches the point to note here is that the brain's "clockwork" is not relevant to the temporal specifications developed or removed in the processes of sequencing and de-sequencing. We can indeed work out what the brain does – the functional specification – without knowing how it does it. This might be disconcerting to some readers, especially neuroscientists and computational neuroscience modellers. Nonetheless that is how we will proceed.

1(2).6.3 There is one aspect of behaviour which is rather surprisingly not addressed by Lashley and his colleagues, nor does it seem to be discussed in more recent accounts of the organization of behaviour. At some scale of specification of action, cognitive representations have to be coupled to the musculo-skeletal system to deliver posture and action. The feedback about the posture and action may be kinaesthetic or more indirectly sensed – perhaps, for example, as sounds from a keyboard or from the vocal apparatus, or in the form of visual inspection of the movement of limbs, or more generally (sound, vision, somatosensory inputs) in respect, say, of control of a vehicle. The problem is that for the most part (that is, other than sequence) the feedback is mediated through channels which have to be "translated" in some sense back to the controlling specification. All this has to be learned.

Animals with precocial emergence, as opposed to species with altricial offspring, can clearly do some things without learning – stand, run – but there are limits even here, and subsequent life-experience/ training can produce behaviour which matches the complexity referred to above.

1(2).6.3.1 The problem, in essence, is calibration of the mapping between specification and effect (and is ignored in behaviourism, and thus in critiques of behaviourism). There are some simple illustrations of the problem. An old fashioned weighing device called a balance is a good place to start. The "scales of justice" image will bring this to mind, if school experience of chemistry lessons (or for some, a visit to a museum) fails. The horizontal beam of the balance can be maintained in that alignment provided the weights in the two pans of the balance are the same. It matters not whether these weights are 1gm each, or 1kilogram each (although the latter might damage the mechanism of a sensitive balance). The load on the centre pivot will be 2 gms, or 2 kilograms – but the beam will remain horizontal.

1(2).6.3.2 The musculoskeletal system delivers for its "owner" (or perhaps "occupant"?) a desired posture or stance, partly by means of muscles in tension across joints. The limbs illustrate this well – hold out an arm or leg and sets of muscles will contract/ relax to achieve the intended result. But by how much will they contract? What is the "resting level" of muscle tension in the limbs to which is added/ subtracted some proportion of tension to achieve the desired posture? Simply inspecting the outcome – a horizontally held arm – tells us nothing about the effort being made: the tensions could be closer to 100grams each, or 1 kilogram each, or 10 kilograms each (as in the case of the balance above where horizontality of the beam tells us nothing about the weights). The resting levels need to be learned, and the control "signals" required to achieve a given posture or action need to be calibrated to the system so that the effect is as desired. For each individual, and from circumstance to circumstance, and with age, and allowing for drug intake and diet, the muscle tensions need to be re-calibrated (this can be done unthinkingly – as when one talks with one's mouth full of food). The Alexander Technique is specifically about this "modelling" of the relation between intention and action. Another simple example, but outside the remit of Alexander, is facial expression. "Put on" a neutral expression and then examine this in a mirror – does the result appear neutral or frowning or smiling in some degree? Remove the mirror and try for an expression which you judge to be a minimal frown, or smile. Re-examine the face in reflection, and you may find yourself surprised by a mismatch between intention and outcome. If so the system needs re-calibration.

1(2).6.3.3 Additional evidence for this complex state of affairs comes from observations of individual differences. People have different gaits, for example, or they move their mouths differently when they smile, or talk – one can become quite distracted watching television news-readers if one examines how they move the mouth (lips, teeth, tongue – see 2(4).8.1.2.7) when speaking. There is nothing fixed or hardwired in these activities. The matching between inner specifications and delivered musculoskeletal outcomes is idiosyncratic because the calibrations are just that – they do not unlock some innate sets of values, they are acquired and adjusted on an individual basis, varied over time and through training, and so on and so forth. Following a stroke, say, significant work has to be done which is loosely thought of as "re-wiring" to provide for effective limb control, or whatever. Likely as not the neurological work is also concerned with re-calibrating the interface between specification and outcome. Similarly, stress induced changes in the musculoskeletal system originate in disruptions to the calibration settings – being tense may indeed be just that.

1(2).6.4 The importance of this observation is that the simple model of the Sequential Imperative, with sequence penetrating the corporeal boundary, doesn't expose the need for further specificational detail (the calibration account, in essence) of how sequences are actually delivered in actions and behaviour. These sequences can be as hidden as vision system saccades, or as obvious as paces across a room. Nonetheless, for sequences to be matched "going out through the skin" and "going in through the skin" there needs to be more complex control of the musculoskeletal system than might at first seem obvious. And, given that the musculoskeletal system is dynamic, and thus itself sequentially managed as behaviour, the implications of the SI are rather more complex than at first appears to be the case (and see the discussion in Chapter 7).

1(2).6.4.1 In respect of GCP2 & 3 we see that these Principles are broad-brush statements which assume a musculoskeletal system, and a sensory system, which can deliver the sequencing and de-sequencing at the finest scale appropriate for components of both action and sensation, and for coupling action intentions to actual muscle contractions. In essence, the issue of componentiality and duality (GCP2) reduces to the requirement that some components are "small enough" to provide appropriate control (or sensory input) which matches or aligns segmentation in action (also in proximal stimuli) with those components – otherwise the management of action (or the success of perception) will be thwarted by mismatched scales. The smallest components in action and perception then link to ever larger scale components "all the way up". Note that GCP3 thus applies at all scales of componentiality – and indeed monitoring works at many relevant scales. We are as aware of the need to speak distinctly in a noisy environment (and we monitor this) as we are of the need to speak sequences of sentences, and as we are of much longer term sequencing in behaviour and perception – the day's events to be planned and reviewed, activities over a year or so, ambitions over a decade – all behaviour is sequenced, and perceptions de-sequenced; segmentation has to be managed at all scales, and simultaneously (see 1(2).5.1.2 and also 3(7).13.1.1).

1(2).6.5 The alert reader may have paused to reflect at this point that of course when learning to perform a complex task an animal needs to acquire a degree of automaticity to attain fluency. Further, when learning speech, say, young humans appear not to analyse fully the sequences of sounds they hear, perhaps revealing that some scales of structural knowledge can be developed later. Children then discover that whole phrases are not unanalyzable entities but are composed of words, for example (see also Peters, 1983). These notions, which relate to the question of segmentation, can be given detailed accounts in the model developed later in Part 2.

1(2).6.6 The GCPs 1–3 discussed here provide some detail of how the Sequential Imperative operates, regardless of the size of brain being considered. The idea really is compelling – it is both very simple and very important. It seems that the brain serves the Sequential Imperative by being a "curious sort of time machine" which manages the exploitation of the time dimension – constructing sequences of fine scale entities for externalisation; deriving from such sequences the atemporal meanings. The reader might wonder if this is linked in any way to the mind-body problem and its solution; this is addressed in Part 3.

1(2).6.6.1 However, some thought should reveal that the Sequential Imperative needs more than just 3 General Cognitive Principles, even though the first three capture the core idea quite effectively. We turn next to elaborate our understanding of the complexity of the SI by considering some more GCPs. Later we will examine in more detail what is meant by "serve" in the statement that the functional specification of the brain is to serve the SI. The next chapter begins with GCP4, and a discussion of learning which, of course, requires changes in stored entities (and even the creation of new ones). This does not undermine the notion that entities are atemporal (GCP2;(i)).

1(2).6.7 The thoughtful reader may have another question in mind at this point. The SI requires bi-directional mapping between sequentiality in action, and atemporal cognitive entities. This has been presented in detail as captured in some General Cognitive Principles (GCPs). A question arises here – are these GCPs necessary for understanding the Sequential Imperative, or merely just one set among many possible sets? My view is that they are necessary, but the possibility that I am wrong doesn't undermine the idea that serving the Sequential Imperative must be the functional specification of the brain. The remaining GCPs may seem less crucial, but again I maintain that the set is required. Without coverage of the ideas captured in the GCPs it would be very difficult to make detailed sense of the Sequential Imperative. The set of GCPs offered here may not be the only way to capture the ideas they capture, but, to repeat, I consider those ideas to be necessarily a part of what has to be said about the functionality of the brain and how it could ever work. The coverage of the remaining GCPs in the next chapter is variable in detail – some material is held over to Part 3. Nonetheless, the totality of the set is presented and the Cognitive Science account is completely sketched out. We will return to this issue in 1(3).7.7.

1(2).6.8 It can be considered that the GCPs on offer in this book could provoke a bigger research project on their plausibility and that of alternatives. That is one sense in which the material provided here can be read as notes for bigger/other projects. There is a different possible project which arises from the material offered above. I have sought to demonstrate that the temporal patterns of activity found via brain imaging of one sort or another have no relation to the temporal patterns in real world activity or in behaviour. The 7Hz pattern identified by Lenneberg (1967; see 1(2).6.2.2 above) is an example of the sort of phenomenon that could be studied in depth.

CHAPTER 3

General Cognitive Principles

1(3).7 In this chapter we look briefly at the remaining General Cognitive Principles and consider how they help to fill out the details of the Sequential Imperative as captured in the first three GCPs, thereby spelling out in some detail the functional specification of the brain as serving the SI. Some of the material may seem a little remote from sequencing as such but we need at this point to set out the scope of the top-down Cognitive Science view of the brain's functional specification in service of the SI. More extended coverage is provided later.

Here are the remaining General Cognitive Principles:

GCP4 Learning serves the sequential imperative.
GCP5 Attention is the management of the processes of sequencing and de-sequencing.
GCP6 Affect is an attentional mechanism.
GCP7 Cognition and affect can be distributed and projected:
- i) in the environment – space, objects, others (not just con-specifics),
- ii) in time – historical, personal.

GCP8 Thought is the production of cognitive entities.

1(3).7.1 In 1980 Chomsky wrote:

> The study of biologically necessary properties of language is a part of natural science: its concern is to determine one aspect of human genetics, namely, the nature of the language faculty. Perhaps the effort is misguided. We might discover that there is no language faculty, but only some general modes of learning applied to language or anything else. If so, then universal grammar in my sense is vacuous, in that its questions will find no answers apart from general cognitive principles.
>
> RULES AND REPRESENTATIONS (1980:29)

GCP4 claims that learning serves the Sequential Imperative; that is why we learn anything at all, and indeed everything we learn. There is no need to posit anything special unless it can be shown that language, say, or patterns of social interaction, cannot be accounted for in the same way as any other aspect of behaviour. In detail, the point of GCP4 is to assert that cognitive entities and

© KONINKLIJKE BRILL NV, LEIDEN, 2017 | DOI 10.1163/9789004342996_006

their componentialities (GCP2, and see Chapter 2 footnote 2) are learned in order for the SI to be served. There is no reason at this stage to suggest that language learning and use requires a special account.

1(3).7.1.1 Learning anything at all pre-supposes successful de-sequencing – perception must yield a segmented input which can be linked to other inputs and to already stored entities. However, successful de-sequencing can be limited de-sequencing. Refer back to Figure 2.2 and the discussion of that diagram in 1(2).5.3.1.3, where it was pointed out that the "surface" elements in the spine are "labelled" for all the different scales of specificational detail. In a creature learning to de-sequence a specific pattern or domain of input, de-sequencing can be successful (i.e. useful) even if the temporal resolution is not yet as finely detailed – fine scale – as it will be eventually; likewise if the complexity of labelling (componentiality) is not yet as thorough as it will become. This issue will be addressed in Chapter 7 but here it is significant because the broad brush conclusion we need now is that learning amounts to development of ever more refined and complex patterns of componentialities; learning is not all or nothing acquisition of the fully specified patterns of componentialties but of ever increasingly more complex, resolved, thorough, accounts of the details "looking both ways" at the duality of the structures (GCP2). We return to this in Part 3.

1(3).7.2 GCP5 should make sense – but then cause concern. The processes of sequencing and de-sequencing need some sort of management; that much seems obvious (else the whole enterprise is determined somehow, which doesn't make sense; or is chaotic, which is unworkable). The idea is that attention provides some sort of control of the relative significance of the various scales of "labelling" of the segments in the spine (consider Figure 2.2 again). But on reflection the reader will see there is a problem, and it is that the attention itself can also be behaviour which has to be sequenced (and de-sequenced for monitoring). In Chapter 7 we will return to the complexities of attention and the need for management of the way the brain serves the Sequential Imperative.

1(3).7.3 Less obviously, GCP6 claims that affect is an attentional mechanism. The issue here is that in addition to attentional behaviour (GCP5) it makes sense to consider attentional states (emotions) and GCP6 is an attempt to capture that insight. That an emotional state may be the (known, desired...) outcome of some other behaviour is not relevant here – what matters is that such states can control sequencing and de-sequencing. The observation does have significance for evolutionary conjectures (think again of homeostasis – perhaps as precursor of emotions in evolutionary terms) and may also be relevant to refinement of concepts of learning. Chapter 7 will provide more detail.

1(3).7.4 GCP7 is more challenging and needs more space here, rather than further referral to a later chapter. Although not explicitly taken up with sequencing and de-sequencing it concerns the management and utilisation of *context* for behaviour. At this point we can discuss briefly the basis for an account of context and contextualization which eventually is required in any case for all behaviour – "situated behaviour" in the current terminology (see Chapters 7 and 9 for further details; on embedded or situated behaviour see also Hutchins, 1995, and Suchman, 1987). The issue we focus on here is that an organism with a brain has to be studied as it behaves and operates in its environment or context; the jargon phrase is "ecological validity" in the study. Abstracting such an organism and making it do things (the rat in the maze, for example), is not ecologically valid in the sense that the context is unreal, and may not even be that well controlled or documented. In Part 3 we will look at context in relation to attention and more complex models of behaviour.

1(3).7.4.1 I have chosen as the point of entry into the complex world of context and contextualization in behaviour the notion of Distributed Cognition (Hutchins, 1995), recently elaborated to cover Projected Cognition (see Edmondson and Beale, 2008). Work on Distributed and Projected Cognition, and thereby on the exploitation of context in behaviour and communication, can serve to ground communication activity, for example, in the cognitive domain generally. As we will see, other behaviour is, of course, contextualized or embedded or situated in the circumstances of its occurrence, but here we will look at human discourse in detail. Discourse might occur in other species, we simply don't know (e.g. dolphins); we do know that behaviour of some species working with humans can look very like communication based Distributed Cognition. For example, control of animals by other animals under human control: sheep-dog working under shepherd control (and this can get quite intricate). Ultimately, context for behaviour may not just be about Distributed or Projected Cognition, or affect, but this is as good a place to start as any (see also Part 3, Chapter 9).

To help the reader focus on the topic the GCP is restated here:

GCP7 Cognition and affect can be distributed and projected:

i) in the environment – space, objects, others (not just con-specifics),
ii) in time – historical, personal.

1(3).7.4.1.1 Work on Distributed Cognition (DC) is not that new – it was a popular topic in the 1990s. At Cognitive Science conferences, and those devoted to Human-Computer Interaction, work derived from or influenced by DC insights was presented and DC concepts were used in analysis of one or another situation, or artefact, or problem (see, for example, Hutchins, 1995; and the work of de Léon, e.g. de Léon, 1999). The core idea is that cognizing has a social dimension, it is not just an intra-cranial activity, but note that this is not an appeal to any sort of distributed mind, or telepathy, or whatever. Note also that although the discussion below is rather human focussed it is possible to sketch accounts of the value of the principle when considering behaviour in a range of species. The GCP actually refers to the *possibility* of distribution and projection of cognition and affect, not the inevitability or pervasiveness (although in fact it is much more widely exploited by humans than might seem possible at first encounter). In addition there is the further point that understanding of the use of context in non-human animal behaviour is not very advanced – we simply don't know how much Distributed or Projected Cognition is going on.

1(3).7.4.1.2 The distribution of cognition – where cognition is here understood quite generally to include memory, planning, perception, reasoning – is found quite widely. Anyone who keeps a diary, either as a journal or as a record of appointments, is distributing their cognition into the environment. Sticky notes attached to the wall near the telephone, with numbers or messages, are distributed cognition, as are shopping lists for the grocery store, birthday reminders on one's smartphone, "to do" lists on one's computer (whether in some application software or attached physically as sticky notes around the screen), and etc.

1(3).7.4.1.3 People working in teams, formally or informally, also display DC in their work. The activity in which they are engaged is inevitably social even if formally structured; there is sharing of activity and mutual understanding – Distributed Cognition – almost to the point that participants feel they "know what the others are thinking". This is readily encountered informally in workshops and kitchens, where one person can "read" the activity of another and engage with it: the need for a specific implement or component is identified and met without a word being exchanged. Sometimes, of course, there are errors; these really show that the attempt is being made to work with DC, rather than to avoid or deny it.

1(3).7.4.1.4 Guide-dogs for the blind provide an illustration of DC; the user of the dog and the dog work as a team to solve cognitive problems. There are occasional reports of pets doing this for other pets – a "seeing-eye cat" for a dog that is blind, for example. Documentation and scientific investigation of such cases is missing, but inter-species helpfulness is probably genuinely reported and need not be surprising. Indeed, keen gardeners know that robins (*Erithacus rubecula*) are attentive to their gardening work, and quite fearlessly perch near freshly turned soil to look for worms/ insects – as if they assumed our work is motivated to help them. It isn't, of course, but the bird's attentiveness is reasonable behaviour in the context of the gardener's work. More obvious inter-species working involves New Caledonian crows (*Corvus moneduloides*) dropping nuts on the roadway for the tyres of passing cars to break open for them.[1] Inter-species behaviour should be studied.

1(3).7.4.1.5 The developmental sequence in the design of implements provides another illustration of the distribution of cognition. An interesting and informative case-study is offered by de Léon (1999) who recounts the development of the firing mechanism for rifles, and the train of thought over many years, and several people, which nonetheless is "readable" or "recoverable" by current technologists reviewing the development of the device (in effect this is a sort of cognitive archaeology).

1(3).7.4.2 The distribution of affect in the same senses as above – in space and time, and involving more than one individual – need not cause surprise. Objects and spaces can be imbued with emotional significance both for individuals and for groups, and over extended periods of time. Attachments can be formed for mating or otherwise and again can extend over long-periods of time, and inter-species attachments are well-known. Attachment can be witnessed by others and thus affect distribution becomes evident. In some circumstances the fact of distributed affect may be evidenced when humans produce special artefacts for one or another purpose (cathedrals, totem poles, photographs in the wallet, etc.); and in some species the permanence of a mate is evident as is the bonding between parent(s) and young.

1(3).7.4.2.1 Fear can also operate in the same range of circumstances – with both humans and other animals capable of associating fear with specific locations, objects or other animals. The association of such an emotion is its distribution, but the fear needs to be witnessed for the distribution to be evident. And of course the demonstration of such affect may be complex; a dog mistreated by one owner may come to fear all humans other than its new carer, and also be relaxed with other dogs – all of which behaviour is displayed simultaneously for all to witness (both other humans and dogs).

1 There are many accounts of these crows – YouTube videos are easy to find. We should probably assume the car is a species for the crow, not a machine occupied by another species.

1(3).7.4.2.2 Territoriality is a mix of both cognition and affect, it would seem, and is observable in many species. It provides an example of distribution both for the self/ owner/ occupant (and see below) and for others (through marking in one form or another). Examples here might include personalizing one's car, or an animal scent-marking its territory.

1(3).7.4.3 The variety of affect and cognition open to distribution does not seem to be limited – examples may require teasing out, and the extent to which it is actually demonstrated will surely vary enormously, both within species and between species. In respect of humans an interesting notion was introduced by Lotman (1990) – the semiosphere.

1(3).7.4.3.1 The semiosphere appears to be the culturally shared milieu, or "space" where affect and cognition are socially distributed and sustained to provide contexts for affect and cognition. "The media" may of course play a rôle, but in general the distributed context is defined and sustained by the participants and events. Thus it becomes clear how such an utterance as "Isn't it terrible?" can work in communication – the semiosphere/ culture provides the reference point(s) but only when assumptions about the interlocutor's location in the semiosphere, so to speak, are valid. When not valid, such assumptions cause confusion and disrupt communication, but that is how context is (or is not) exploited. Indeed, the exploitation of context – contextualization – is the key idea; context itself is indefinitely variable.

1(3).7.4.3.2 Contextualisation is the invocation of distributed cognition and affect, both by the creature who produces the behaviour in context, and by the creature who "reads" that behaviour by reference to recoverable context. There is a difficulty, of course, which is that the distribution may be overwhelmingly idiosyncratic and/or fleeting leaving the observer, or recipient, or interlocutor (the term varies according to the nature of the creatures involved, and the behaviour) perhaps only aware that contextualisation is operating, but not otherwise able to recover any sensible/ convincing/ reliable interpretation. Such distribution causes difficulties for others, but is linked to a more private form of what looks like Distributed Cognition but is more sensibly called Projected Cognition. We turn to this next, before returning to look at problematic DC.

1(3).7.4.4 Projected Cognition and affect are the private equivalents of the more public distributed phenomena (and see Edmondson and Beale, 2008). The projected forms are of course distributed into the environment, but without advertisement, as it were. The knot in the handkerchief, the fear and avoidance of a particular location or route (having once witnessed something unpleasant) which is not even noticed by anyone else; these and many such personal preferences in behaviour are projections into the environment of cognition and affect, but only really intelligible to the one who projects. Sometimes the projections become less "of the moment", or less intensely private, and more enduring and shareable and gain status as distributed cognition/ affect. An example here would be the use of a particular tool (say an old-fashioned bicycle tyre lever) to open paint tins; the initial such use is projection, the habitual usage is distribution. The ability of a human (typically, but not exclusively) to recognize in an object its value for some function and then to recast, as it were, the object in that functional rôle, is simply the ability to project cognition. This may remain private and idiosyncratic, but where it leads to a shared understanding of the new rôle for the object one can speak of the cognition being distributed – and acquiring a sort of permanence or inflexibility. Primates who fish for termites by fashioning twigs into dipping rods, or who break open nuts with stones, project onto those objects their cognition. This is observable and we humans who observe it might consider the outcome to be Distributed Cognition. Cultural transmission of the technique is not uniformly, or perhaps even convincingly, demonstrated; thus although we can see what is going on (or think we can – anthropomorphisation is licensed in this account), another con-specific primate may not understand either the problem or the solution in the same way, and therefore the projection remains essentially private. But even where there seems to be compelling evidence for cultural transmission of tool making behaviour, as when chimpanzee mothers appear to instruct or shape the behaviour of their offspring, alternative accounts can be offered (see the discussions in Tomasello, 1990, and 1999: 26ff.).[2]

1(3).7.4.5 Contextualisation in human discourse may involve both distribution and projection of cognition and affect. Topic selection and social reference along with the deployment of appropriate pragmatic techniques, including

2 There is an important general point here – apparent uniformity of performance may mask heterogeneity in the population. Individual experiences – a necessary fact of life – can lead to underlying differences in cognition and skill even when the observable performances look the same. We will encounter this heterogeneity again in relation to speech production (Part 2) and more generally in relation to language learning (Part 3). For completeness here it should be noted that there is some evidence from attempts to teach young adult humans to make stone tools that some people never seem to be able to do it (Ohnuma et al., 1997). The sample studied was small, but the outcome was nonetheless provocative. Making stone tools is difficult!

gesture, expression and eye gaze, can be essentially public – Distributed Cognition. Such distribution is indeed expected and makes the discourse less pedantic. That's how context works. And of course for both participants the exploitation of context is part of the (de-) sequencing activity in language which goes beyond the stream of sound to reference locality and other actors. In situations not involving humans, or human discourse, context still works the same way – behaviour is contextualized through Distributed and Projected Cognition (see below: 1(3).7.4.6).

1(3).7.4.5.1 Discourse becomes difficult, and may indeed break down, when the cognition and affect are projected and essentially private – context is not recoverable by the other participant(s). This can be observed when a "private joke" is shared between two people in the presence of a third for whom the references are not recoverable. The cognition and affect are both simultaneously distributed (between the two who share the joke) and projected (i.e. where the details are hidden from the third who is only aware that something is "going on"). It can happen that the third conversant is unaware anything is going on at all – unaware that there is any sort of context to be recovered.

1(3).7.4.5.2 Contextualization is deployed pragmatically for management of conversations – topic shift can be accomplished by more or less explicit adjustment of contextualization. Sometimes one's conversant is clearly contextualising but the result is so idiosyncratic as to render the recovery of meaning unlikely – and in such cases the conversation may move into a repair phase (or not, and the listener, more likely, is left wondering what was going on). In these cases the attempted distribution of cognition has failed, and the listener is left knowing that something was projected but not what it was.

1(3).7.4.6 As already noted contextualization is not all about human discourse, although that does seem an easy domain in which to elaborate some of the concepts. Any creature walking on an uneven surface, or climbing up a rockface, or swimming in the sea, or whatever – or perhaps observing another creature doing such things – has to reconcile or negotiate the facts of the environment in relation to the activity (or indeed observe this process). This is contextualization. Sheep train their young how to navigate and use mountains and pasture both for effective grazing and also for control over straying. Such learning about their environment (called hefting) contextualizes their behaviour. A ewe "taken out of context" is lost (for a while) just as phrases or words taken out of context can change their meaning. There is more discussion of GCP7 in Part 3, where we find that attention manages contextualization and its readability by other beings.

1(3).7.5 Our last General Cognitive Principle for consideration here is also obscure, but by virtue of concision instead of strangeness. GCP8 has it that "thought is the production of cognitive entities". This needs unpacking/ elucidating – and then in a later chapter linking back to the Sequential Imperative.

1(3).7.5.1 Cognitive entities have a Janus-like componentiality – they have as components other cognitive entities, and they are in their turn the components of other entities (GCP2;(ii)). What is not covered in this structural account is that cognitive entities also have content, values, substance... – whatever we wish to call it this content is more than just the indexation for the connectivity in the mesh of components and sub-components. Thus the production of cognitive entities can mean both the creation of new content and/or the creation of new links in the mesh of entities whereby an entity gains sub-components or becomes a sub-component. The cases for consideration here are thus (a) the creation of a new node in the mesh of components, complete with its links, and (b) the creation of new links in the mesh for an established node (which might thereby undergo some change in its content).

However, the term *production* is intentionally a little vague – there is a need to cover the use and deployment of entities in the grand scheme of behaving on the one hand, and perceiving or learning on the other. Which is to say, cognitive entities are *for* the functioning of the Sequential Imperative – one thinks in order to serve the SI (yes, really!). The brain will serve the SI in a number of ways, as we will see later (Chapter 9), but here it suffices that one's appreciation of the creation of cognitive entities (thought) comes through increasing the potential for, and variety of, behaviour – the entities influence sequencing, even to the point of producing sequences of behaviour (speech, for example) and its internal precursors. We will also see (Chapter 9) that cognitive entities can be created via de-sequencing. This is not just learning – there are creative ways of seeing aspects of the world which "reveal" previously unrecognized aspects of events or objects. This counteracts the previously noted bleakness which seems to be part of understanding the Sequential Imperative (1(1).4.7).

1(3).7.5.2 Production, so far as we need to develop the concept at this stage, is thus creation, modification, and deployment of cognitive entities both within the overall interconnected (and developing) mesh of such entities and also within the apparatus for sequencing/ de-sequencing which we can envisage as delivering the Sequential Imperative for the body with the brain.

1(3).7.6 The Sequential Imperative, it has to be remembered, is an imperative which unavoidably couples the brain of an organism to its behaviour and its environment. But, and this is crucial, the reality of the Sequential Imperative is only the starting point. The functional specification of the brain is to serve the SI. There are two senses in which the verb "serve" is being used – "support", or "makes possible behaviour" as in GCP3, and "deliver", which is our focus in the next Part. The "support" rôle requires that the brain stores entities (at all scales) in atemporal form for sequencing and in response to de-sequencing of inputs (and as a consequence of learning (GCP4) and of thought (GCP8)). The "delivery" rôle concerns the mechanism which does the (de-) sequencing. Our answer to the question posed at the outset – "What does the brain do?" – now has two separate aspects: the creation/ storage of the entities, and the use of the entities. In respect of the use of the entities, we see that the brain must provide a mechanism of some sort which delivers the (de-) sequencing in ways which exploit as much as reflect the General Cognitive Principles, and to discuss this we need a model. Such a model has been sketched out earlier, in Part 1(2).5.3.1. Part 2 provides a much more comprehensive model.

1(3).7.7 Part 1 has focussed on the big Cognitive Science idea, and set out the range of issues which need to be considered when reflecting on the Sequential Imperative. The reader should be in no doubt at this stage that the SI is blindingly plausible – obvious, even. The functional specification of the brain – any brain, anywhere, any species, any time – must be to serve the SI. We have explored the need for General Cognitive Principles as both encapsulating the SI and showing what is required to serve the SI. This exploration is incomplete thus far, but has covered both a non-hierarchical scheme for structuring both articulatory output and sensory input, and a discussion of what is meant by "atemporal" in respect of cognitive entities (addressed again in Part 3). The set of GCPs is open to change and refinement but the requirement for some such set, and one which covers the content identified by the current set, seems less in doubt. The reader might find the first three GCPs more compelling as a paraphrase of the whole notion of the SI. The remaining 5 GCPs could seem underwhelming, and in any case the treatment in Chapter 3 is somewhat sketchy – except for GCP7. The point being made here, however, is that all the points captured in those GCPs need to be addressed somehow if the SI is to be made workable. They cover the implications, as it were, of turning the core idea into structured behaviour – in all species with brains.

1(3).7.8 It is important that the reader is comfortable with the notion of the Sequential Imperative and with the idea that the functional specification of the brain is to serve the SI. The next major Part – 2 – turns to consider in detail a model apparatus or mechanism which can be envisaged as a functional model for *how* the SI is served by a brain – any brain.

The model[3] is derived from a version of phonology, but this does not mean that all creatures with brains can speak. Rather, it means that the organization of speech, as captured in a specific formalism, reflects the organization and utilization of cognitive entities more generally, and necessarily as required for successful delivery of (de-) sequencing. This model/ account/ apparatus is derived from a detailed account in phonology and is thus grounded in understanding of speech behaviour; to understand the model the reader needs to understand quite a bit about speech. The details can seem intimidating but really do need to be understood to appreciate the complexity of delivering the sequencing and de-sequencing. This complexity is addressed again in Part 3, in Chapter 7. Much of the discussion focusses on the notion of the syllable in speech, and this is done in part because some cognitive scientists consider the syllable to be another core idea (as we will see in Part 3), and in part to help the reader form an intuitive appreciation of the cognitive generalities being discussed.

1(3).7.8.1 Your author has an additional task at this point in the book. The more general work now needs to be shown to be relevant to understanding how language works. The use of the diagrams in 1(2).5 and the discussion of contextualization already begin the blending of Linguistics with more general cognition – but the question remains for the reader: does the account of Cognitive Science sketched out in Part 1 cover the possibility of language behaviour? In one sense the answer – affirmative – is already clear. The Sequential Imperative is obviously unavoidable. It covers any behaviour, including language, as it is externalised and internalised. It must do so, irrespective of explanations about what happens in the cognitive hinterland. What is left then is to argue either that language uniquely has some other *additional* structuring principles which cannot be accounted for by the SI – but which nevertheless must comply with the demands of the SI (without disruption). Or, that language is exactly like other behaviour in the general sense – and in Chomksy's own words (1(3).7.1) this would render UG vacuous. If the latter is to be contemplated then probably more work needs to be done on showing that all of language (including learning) can be covered by the SI in detail and that innatism has no value (but see Chapter 7). This quest would go beyond the sketch offered in this book, and thus is another, bigger, project, indeed perhaps several projects! However, it may be that some readers find that even the sketch offered here is compelling and that there is no need for the bigger projects. Note again, whatever story is told about language the actual production and perception must conform to the SI, that is unavoidable.

3 The model does not attempt a *neurophysiological* approach to how the Sequential Imperative is served.

1(3).7.8.2 Part 2 will also answer the question affirmatively, looking up at the Cognitive Science from the linguistic details. The answer is developed by exploring a model derived from language behaviour and showing that this serves the Sequential Imperative. Indeed, the Linguistics insight with general Cognitive Science significance – the deployment of a linearization model from phonology for the delivery of the SI – not only reaffirms the overall Cognitive Science approach, but neatly removes the need for any account of language behaviour as special or privileged in any way. The material in Part 2 is necessarily rather complicated and for those who have not thought about the detailed structure of language it may seem like too much hard work. However, the effort is worthwhile because the outcome has surprising generality and wide-ranging implications for understanding the evolution of brains (as we will see in Part 3).

1(3).7.9 Chapter 4, the next chapter, provides a general overview of linguistic data which Cognitive Scientists should know, and be prepared to embrace as part of the possibility of language they must demonstrate in Cognitive Science. The survey demonstrates the complexity of speech, let alone language. Chapter 5 goes into perhaps overwhelming detail – necessary to show what work has to be done in sequencing and de-sequencing for speech use, and also how this can be modelled. It also reveals that learning a language, or several (as many children do), involves a huge amount of detail much of which is not obvious to the child, or eventually the adult. There is some discussion of what has to be learned by a child, in the sense of quantity and complexity of linguistic detail, and in the sense of the use (or not) of rules. (This is interesting in its own right, but is particularly relevant for subsequent reappraisal of GCP4 in Chapter 7.) Chapter 6 provides some information about a computational version of the functional model (this is not a computational model of the brain or part thereof). It will become clear just how it is that understanding speech can indeed lead to a model for all behaviour.

1(3).7.9.1 Homeostasis is not our focus in the book, as noted earlier, and the topic is set aside in Part 2. The idea worth working on as a bigger project is that affect is a state-like mode of attention in (de-) sequencing, which may have emerged as the (de-) sequencing engine evolved in complexity and the opportunities for homeostasis became more complex (see Part 3 (3(7).13.6.3)).

1(3).7.9.2 There are two related loose ends in the material presented thus far, and not yet flagged for attention later. The first concerns atemporality, and the second concerns the mind and conscious experience. These can be thought of as two sides of the same coin.

The issue really goes back to 1(1).4.7.1 and the exhortation there that the reader needs to be sure they "get" the idea of the Sequential Imperative. Conscious reflection on experience of, say, having a belief suggests that this feels unlike the experience of, say, making a judgement. Beliefs can be accepted as atemporal in experience; judgements feel like a temporally organized process. So, doesn't this undermine the notion of ubiquity of atemporal cognitive entities?

No, because the atemporal nature of cognitive entities is a property of the brain, not the experiences. And, as has already been mentioned (and will be again, in Part 3), there are processes that will be changing/ creating cognitive entities (e.g. learning) in the brain which are not temporally aligned with the external world. These activities are temporal in a sense which doesn't matter in relation to the Sequential Imperative, and which in any case may not be what we think we are aware of in conscious experience.

An illustration of misplaced appreciation of temporality arises in consideration of arithmetical notation. Simple formulae, such as 3 x 4 + 2, are ambiguous as written but can be clarified with additional notation: 3 x (4 + 2) and (3 x 4) + 2. These formulae have a process feel to them. The arithmetic reality is that the formulae:

$$3 \times (4+2) = 18 \qquad \& \qquad (3 \times 4) + 2 = 14$$

are statements of identity, not process. We are deluded into a process experience by the conventions of processing the notation as we read it; the reality is atemporal. My point is that much of experience has this characteristic (because it is linguistically re-presented to us as inner speech, which is necessarily sequential). Where there may indeed be some temporality in the creation of cognitive entities, as per GCP4 and GCP8, that will not necessarily be accessible to experience. These mental processes are, in a phrase which crops up later, "off-line" and decoupled from the flow of time in which the creature with the brain is embedded. The point of the Sequential Imperative is that the brain has to traffic cognitive entities into and out of sequence, and to a first approximation it makes sense here to say that the cognitive entities are inherently atemporal. A more nuanced account, in Part 3, explores the detail but for now the reader is encouraged to set aside their experiences and focus on the functionality of the brain. Part 3 also considers the case against innate knowledge (of language). It needs to be repeated that we are considering all creatures with brains, not just humans.

PART 2

Serving the Sequential Imperative

Introduction to Part 2

In this Part we consider the delivery or realisation of the Sequential Imperative in an organism with a brain. The SI is universal but the organisms are not. The delivery of the SI thus needs to be thought of as some universal scheme which is realised in particular ways, species by species, but underlyingly all the same. The need is for a sort of "scale independent mechanism" in the brain, so to say, which delivers the SI for an organism – specific for each species, but nonetheless an instance of the general mechanism. Additionally, we need to keep in mind that this mechanism must evolve along with the organism; eventually we must provide some sort of account of how this might be understood (see Chapter 8). And note – we are not concerned just with humans, and we are not concerned with the neurological details.

I will be presenting as completely general a structural account derived from a uniquely human activity – speech. This is, it could be said, quite a stretch, but I believe the reader will find it plausible. And yes, I am aware of the difficulties this will raise for some. In the mid-1970s I was working on developing a speech training aid for deaf children, and in consequence found myself visiting schools for the deaf in the London area. The prevailing educational "philosophy" at that time in UK was that signed language was not appropriate – the various different signed languages around the world were thought by some not to be real languages, and thus for advocates of signing to propose the adoption of one such "language" for use in classrooms was thought by many educators to be unfair to the children, who in any case had to learn to speak (somehow). I met the head teacher of one school for the deaf who told me forcefully that speech differentiated humans from apes, and that gesturing was for apes – the children had to be taught to speak to demonstrate that they were human. This is, of course, nonsense. Signed languages are not to be thought, by the reader, as excluded by the focus here on speech – the focus on speech is just because speech is the most studied of all animal behaviour. Contrariwise there are, without doubt, many people who believe that uniquely human behaviour (language, music, graphic art, dance…) requires a unique account which cannot apply to other species. This too, I suggest, is nonsense.

Interestingly, the general structural account derived from speech serves well for developing accounts of signed language morphology and phonology. Despite not being based on sound, signed language phonology is the term used to deal with the gestural equivalent of spoken language phonology – and this already points the way to surprising generalisation: amodal characterisations of linguistic activity are readily undertaken (see Meier et al., 2002). So, by

© KONINKLIJKE BRILL NV, LEIDEN, 2017 | DOI 10.1163/9789004342996_007

referring to the uniquely human activity of speech I do not mean to downplay the significance of signed languages, both for Deaf communities and for the field of Linguistics. Likewise, it is not the case that I am somehow seeking to undermine the uniquely human in humanity. I focus on speech because the data lead me there. Understanding the most studied of all behaviour is, I believe, key to understanding all other behaviour.

The focus in this Part is on speech and language and the reader needs to understand that this illustrates, perhaps in extreme form, the general way in which the brain serves the Sequential Imperative. The behaviour of other creatures with brains must fit the story which unfolds but the perhaps extravagant concern with minute details in speech may mask, for some, the generality of the account. The bigger picture here is of speech and language providing the detail from below to match the Cognitive Science account in the first Part. The reader will form the impression, likely enough, that the details are impressive, but how can it all work? The value of speech as an example is we know it does work. The next Part takes up the challenge of explaining how the complexities of behaviour, including speech, are managed.

This Part comprises three chapters, in the first of which we discuss some of the intricacies of human speech and language. There is a lot of potentially intimidating detail. The middle chapter presents a model of the organisation of speech activity and shows how this can be generalised to all activity, and in a scale independent way – and there is more detail. The third chapter considers the implications of the model when viewed as the "mechanism" for serving the Sequential Imperative. The reader will find that 1(2).5.3.1 provides a very brief introduction to the material to be offered in detail in this Part.

This Part provides the linguistic detail and model needed to feed into the Cognitive Science to make it more general and capable of covering speech and language. The perspective is thus complementary to the top-down Cognitive Science discussed in the previous Part. At first the complementarity will seem disjunctive and obscure rather than helpful. However, the reader is urged to ask themselves from time to time – if the Sequential Imperative has to deliver all this complex structuring of behaviour for language, perhaps it isn't unreasonable to think that language is the right basis for modelling all behaviour?

CHAPTER 4

Structure in Language

2(4).8 We commence our brief survey of structure in language with a look at speech production. The account is brief because many linguistic textbooks cover the material in detail and there is little point in inserting here a shelf-load of textbooks (on speech the reader is recommended to look at Ladefoged and Johnson, 2011; and Hayes, 2009). Additionally, there is a great deal of research being done on speech production and perception, and systems of explanation are not yet unified – much remains to be done. Nonetheless enough is known, and in detail, to serve our purposes in this Cognitive Science adventure. Speech is our focus simply because it is so well studied, but in addition there are exercises the reader can try at home, and this effort is more readily accomplished than if the examples are, say, manual gestures. After considering the possibly surprising complexities in speech production we move on to look (very briefly) at morphology and other structural issues. Our concern is to reveal something of the extent of complexity and arbitrariness in human language behaviour, all of which has to be learned (see also Comrie, 1989). The next chapter provides more structural detail.

2(4).8.1 The vocal apparatus in humans is capable of making quite a large range of sounds for use in speech; probably several hundred distinct sounds. No one language uses the full set of all possible sounds. Different languages use different sub-sets of the range of possible sounds and each sub-set is systematically organized rather than randomly selected – with some subtle additional factors involved. The complexity at this scale of structure seems surprising to many people. Because the complexity of the model discussed later reflects some of this complexity in speech production we need to take some space here to cover the details (but not, to repeat, at the density to be found in textbooks or research volumes).[1]

2(4).8.1.1 Speech appears to be made up of sequences of "segments" or speech sounds (we call them consonants and vowels) – and each speaker's task is to

1 The reader may have encountered the "Message Model" – a widely assumed approach to modelling communication (e.g. see Lotman, 1990; Hauser, 1996). The "Message Model" has it that communication involves the production of coded messages, their transmission via a noisy channel, followed by reception and decoding by the recipient. It is not part of the story I tell; in my view it is worthless.

© KONINKLIJKE BRILL NV, LEIDEN, 2017 | DOI 10.1163/9789004342996_008

arrange the sounds in the right order to express the words (sentences, thoughts…) that they want to utter. We should take some time to unpick this conception of speech and look at fine details, assumptions, and data. There is a gentle introduction to this topic below, with more detail in the next chapter.

2(4).8.1.2 There is an important distinction to be made between the production of speech and the hearing of speech. Indeed these two aspects are different to the extent that they give rise to research in two domains – articulatory phonetics and perceptual phonetics. There is a third domain – acoustic phonetics – which is concerned with the evidence produced by the articulatory activity and subsequently processed in perception and it is convenient (here) to align acoustic phonetics with perception. To be sure, the simple conception of two domains – acoustic phonetics aligned with speech reception, and articulatory phonetics with speech production – is a little forced, but it suffices for now. We will concern ourselves mostly with articulatory phonetics here. Pursuing our concern with arranging the production of the desired sequence of speech sounds we note that this becomes the issue of arranging the desired sequence of articulations. The sub-set of possible speech sounds (see above), for any one language, thus becomes a set of (idealised) articulations for that language. If the reader at this point starts to wonder if the sounds are somehow more clearly identified than the articulations are specified, then they are on the right track. One of the complexities we need to consider in the production of speech is precisely this issue – different articulatory work can produce a given speech sound, so far as a listener is concerned (the acoustic measurement may say otherwise, and therein lies the difference between acoustic phonetics and perceptual phonetics – including categorical perception (see 2(4).8.3.2)).

2(4).8.1.2.1 Let us consider a few examples and thereby dive into some detail. This may be a little intimidating; speech really is very complex to produce. Consider first the issue of silences. Text is really very misleading – we do not speak with gaps between words in the sense reflected in the text you are reading. Indeed in some languages the written form can consist of character strings without gaps between the words (e.g. Thai). The hearing of the continuous speech stream, and the successful comprehension of what is said, permits the listener to inject the word boundaries into the sound stream; they are not produced *per se*. In this way the casual listener's conception of speech as a sequence of words is an overlay of perceptual effect on to the continuous sound-stream.

2(4).8.1.2.2 How continuous is the stream of sound in reality? Apart from pauses and the "ahems", "ahs", "ers", etc. of normal speech, are there really no gaps?

The sense of gaps can arise as a consequence of the natural alternation between sounds involving the vocal folds – the sound production mechanism in the larynx – and those which rely instead on turbulence in the airflow through the mouth (setting aside for the moment speech sounds involving both sources of sound).

So, for example, a sentence:

> Cats are quiet when stalking prey.

sounds "gappy" when contrasted with the sentence:

> Why were you away a year, Roy?

and indeed this difference can be felt if you place a hand lightly on the throat as you slowly speak each of the two sentences. The second of the two is uttered without break in the activity of the vocal folds. But the breaks in the voicing, as it is called, in the first sentence are not produced and heard as silences, and the very brief silences detectable instrumentally are not uniformly aligned with the word boundaries, as we will now explore.

2(4).8.1.2.3 There are indeed some very short "silences" in the first sentence – for example in the 't' in 'cats', 'quiet' and 'stalking', before the 'q' in 'quiet', the 'k' in 'stalking', and the 'p' in 'prey'. These sounds involve closing the vocal tract – the air-passage through the mouth – in order for the distinctive character of the sound to be expressed by the release of air (creating turbulence which produces noise) as the stoppage is "unblocked". Indeed, the reader is encouraged to try artificially lengthening each of those stoppages by holding the tract in the stopped configuration for a few seconds before releasing the air and creating the desired speech sounds. In normal speech the "silences" are very brief indeed, and are not really perceived as silences – and they don't mark the word boundaries.

2(4).8.1.2.4 Yet more complexity is revealed when one of these silences is explored in more detail. We focus now on the sounds 'p', 't', 'k' in two different contexts. We can exemplify these contexts with the English word pairs: 'pin', 'spin'; 'tone', 'stone'; 'can', 'scan'. To the untutored ear, and the monoglot English speaker, the consonants 'p', 't', 'k' in the three pairs of words sound the same. To speakers of some other languages the 'p' in 'pin', and the 'p' in 'spin', for example, are two different speech sounds. And indeed they are in fact differently produced by speakers of English – as is readily demonstrated. If you speak each of the six words, with an open hand held with the palm close to the mouth, you will notice that the first of each pair is accompanied by a brief puff of air (a sheet of paper dangled in front of the face does as well – moving with the puffs, but not otherwise). The effect is most noticeable with the 'p' sound. The sounds with the air puff are called *aspirated.* (Holding a microphone too close to the mouth when using an announcement system typically produces popping noises with the aspirated sounds, but not the other sounds).

To cover this point with appropriate technical detail, and formalism, we can do the following. We note, initially, that the linguistic specification of the consonant we call (in English) 'p' is called the *phoneme*. This is normally written /p/. The actual sound, usually called a *phone*, is normally written thus: [p]. This phone, however, is only one of the two realisations discussed above. The aspirated phone is usually written [p^h]. These two different *phones* are referred to as *allophones* and they occur in specific situations – by specification, not by choice. Where the [p] is used the other is not, and vice versa (so we get [p^hin] and [spin]); the phonologist calls this "complementary distribution". Speakers of English know this, and stick to the "rule" for use of the specific allophones. But they usually don't know they know about the aspirated form or its restricted use. And native speakers of English will have learned the two speech sounds and their appropriate usage without becoming aware of the differences.

2(4).8.1.2.5 Technically, the difference between the two allophones of /p/ is a difference in voice onset time (VOT). VOT is the time difference (positive for a delay) between the release of the stopped/ blocked oral tract and the onset of vibration in the vocal folds in the larynx which sound the next phone.[2] The VOT for the unaspirated allophone in English is ~15 milliseconds, compared with ~55 milliseconds for the aspirated [p^h] form. In other languages the differences will be different (see Ladefoged and Johnson, 2011). An unusual case is Navajo where the [g] has a VOT of ~45 milliseconds, and the [k] (without allophonic variation) has a VOT of ~150 milliseconds. Note here that VOT is not a property of the vocal tract – it is arbitrarily controlled/ managed for different languages and sometimes serves to differentiate meanings, sometimes not.

2(4).8.1.2.6 It is sometimes assumed that the child faced with learning English learns just to discriminate those sounds of English which matter for differentiating meanings. This is an odd assumption, in fact, because of course children

2 In passing, the reader should note that VOT can be negative, and that voiced stop consonants can be produced with a breathy characteristic (see Ladefoged and Johnson, 2011:155) – all of which increases the range of sounds to be learned and used by speakers of human languages.

learn speech mannerisms and other aspects of speech production as well. Actually the task facing the learner is better described as learning all that goes into speech and *in addition* learning which activities are important for meaning differentiation.

The child learning a language, and to speak it, has to unpack the complexities in the sounds she hears, and perhaps through experimentation (and feedback from others) she learns what she has to do with the vocal tract (and speech related musculature) to produce acceptable equivalents. The vocal tract size differences, and the differences between the pitch of the voice in men, women and other children, all provide evidence of irrelevant differences, and idiosyncratic aspects provide for speaker differentiation and identification. But still the child has to explore the range of activities required to modify the vocal tract in the manner(s) required to produce sounds recognized as speech (or speech relevant). Interestingly, the child also learns to control such things as VOT with sufficient precision to produce the allophonic variations mentioned above, even though when quizzed native English speakers are not aware of the difference or that they control their speech that precisely. And, to repeat, the difference is not meaning bearing (in English), even though it is a part of the language rather than being a personal idiosyncracy of production.[3]

2(4).8.1.2.7 The production of the /p/ which has been our focus here is described in Ladefoged and Johnson (2011:278) in some considerable detail for [p]:

> Underlying our linguistic description of [p], to take one simple sound as an example of speech motor control, is a dizzying array of muscular complexity involving dozens of muscles in the chest, abdomen, larynx, tongue, throat, and face. And all of these must be contracted with varying degrees of tension in specific sequence and duration of contraction. [*Some of the details are spelt out.*] …
>
> So coordination of the four main lower lip muscles is complicated and can't be specified with predetermined target "tension" levels because the actual degree of muscle fiber activation for raising the lower lip in [p] depends on the tension of the other lip muscles. But the situation is even more complex than this because the lower lip moves up and down as the jaw moves up and down. [*The details involving another six muscles are spelt out.*]

3 Actually it is the scheme of *contrasts* developed or learnt by the child which matters. See 2(4).8.3.2.

Rather than go into further anatomical detail of other muscle groups and other parts of the body involved in bringing the speech gestural target [p] to successful delivery (and remember – all this is transient as the sequence of vocal tract gestures unfolds in speech), Ladefoged and Johnson (op. cit.) go on to consider the wider issues of equivalences and variation. A speaker can, for example, deliver sequences of speech gesture targets (articulatory configuration targets) more or less successfully (but usually enough for intelligibility) when their mouth is constrained in some way (holding the teeth clenched, or biting on a coin on edge to keep the jaw a little bit open but fixed). The speaker compensates for the constriction by "recomputing" the instructions to some of the "dizzying array" of muscles (ventriloquists learn to do this). Likewise, speakers have so much individual freedom in how they actually manage the "dizzying array" that they end up with a private interpretation of what is required to deliver gestural configurations/ targets recognizable as phones. Indeed, this is readily observed if you scrutinize the faces of news-readers on television (see 1(2).6.3.3). Some clear idiosyncracies are observable in lip and jaw movements, and sometimes also tongue movements. The configurational/gestural targets are being achieved *well enough* to produce recognizable speech (assisted by categorical perception – see 2(4).8.3.2).

2(4).8.1.2.8 Once the child has learned how to configure the vocal tract (and the dizzying array of musculature) to produce the systematically contrasting sounds which work in linguistic communication, whilst at the same time learning what idiosyncratic vocal tract activity doesn't much matter, she has to go on to learn and exploit other aspects of the sound production system – e.g. variations in pitch and rhythmic patterns of stress and emphasis – some of which are meaning bearing, and some just part of the structure of the language (and arbitrary).[4] This leads us to consider another speech "unit" – the syllable.

2(4).8.1.2.9 The aspirated allophones in English are used in syllable initial position where the syllable is stressed in the word.[5] So the *pin* and *spin* difference emerges because in the second word the /p/ is not syllable initial, even though

4 Control of breathing is also managed for speech, and this has to be learned.

5 The account here is simplified! Kahn (1976) argues that the distinction between aspirated and unaspirated stop consonants /p/, /t/, /k/ is actually just that the aspirated forms occur in syllable initial position. The degree of VOT delay signalling aspiration is merely largest in the case of stressed syllables where the consonant is syllable initial, but always larger than the unaspirated VOT for those consonants in non-syllable-initial locations. The issue is one of categorical perception, not acoustic measurement; although the latter is required to detect the differences it does not provide a definitional criterion. Ladefoged and Johnson (2011:153) do not discuss the intermediate values of VOT perceived as aspirated.

the syllable (word) can be stressed. In a word such as *purpose* the two allophones are used, the first /p/ being aspirated, the second unaspirated. In isolated citation form, the noun version of *import* has /p/ in the unaspirated form, whereas in the verb the same word will have a stressed second syllable, and hence the /p/ will be aspirated (stress is complicated in English because other stressing in a sentence may over-ride the apparently necessary stress patterns in word forms – see Ladefoged and Johnson, 2011:250). Speakers learning a language have to learn the way syllables carry rhythm and stress patterning for that language – and languages differ in this domain as much as any other. In their discussion of syllables and stress Ladefoged and Johnson put it this way (2011:249/250):

> English and other Germanic languages make far more use of differences in stress than do most of the languages of the world, having somewhat variable word stress so that the location of stress is not always predictable from the segmental structure of the word, for example, (*to*) *insult* versus (*an*) *insult*, or *below* versus *billow*, or *market* versus *Marquette*. In many other languages, the position of the stress is fixed in relation to the word. Czech words nearly always have the stress on the first syllable, irrespective of the number of syllables in the word. In Polish and Swahili, the stress is usually on the penultimate syllable.
>
> Variations in the use of stress cause different languages to have different rhythms, but stress is only one factor in causing rhythmic differences. Because it can appear to be a major factor, it used to be said that some languages (such as French) could be called syllable-timed languages, in which syllables tend to recur at regular intervals of time. In contrast, English and other Germanic languages were called stress-timed in that stresses were said to be the dominant feature of the rhythmic timing. We now know that this is not true. In contemporary French, there are often strong stresses breaking the rhythm of a sentence. In English, the rhythm of a sentence depends on several interacting factors, not just stress. Perhaps a better way of describing stress differences among languages would be to divide languages into those that have *variable word stress* (such as English and German), those that have *fixed word stress* (such as Czech, Polish and Swahili), and those that have *fixed phrase stress* (such as French).

2(4).8.2 The definition of a syllable is not straightforward, but speaker intuitions are remarkably robust, which means that discussion of syllables need not be difficult. There is further structural elaboration of the concept of syllable in the next chapter. We need to remember here that all speech is delivered syllabically (necessarily so), but that does not mean that speakers are aware of syllables, or that they deploy them in a linguistically structured manner. Some languages are self-evidently organized syllabically (perhaps in the sense that a syllabic writing system is used)[6] but use of the syllable as a unit in linguistic structuring in that language may not be so obvious (concerning Japanese see Kawahara, 2016). This is an example of the concept of semiotic freedom noted in GCP1;Cor2.

2(4).8.2.1 The structure of a syllable can be approached in two ways – sonority and constituency. Working with sonority as an organizing principle means describing in a systematic way the quality of the sound produced by the vocal tract in a range of configurations and showing how that quality varies during the production of a syllable. Put simply the idea is that some sounds are more sonorous than others (more resonant, not just seeming to be louder), and further that a syllable is always a sequence of such sounds arranged to provide a pattern in the sonority variation where the core, or nucleus, of the syllable is the sonority peak and the values for sonority fall away either side to a minimum between syllables. It emerges naturally from such a perspective that the syllable is in some senses the smallest unit of speech production (see 2(5).10.3.4.2). Furthermore, other noises from the vocal tract (e.g. coughs) do not demonstrate a sonority contour of the required shape and thus are not heard as speech (they are not syllabic). The appeal of such an account is that it makes sense of language independent intuitions about syllables in speech – how to count them, make them rhyme, and so on. Additionally, on those occasions when a consonant serves as a syllable nucleus (the *n* in the utterance *fish'n'chips*, or *cut'n'paste*) a sonority viewpoint makes sense (the *n* is a local sonority peak). A difficulty is that the scheme is imprecise about boundaries between syllables and cannot explain some of the details found in typical syllabaries, such as that all the entries are either isolated vowels or a consonant followed by a vowel (but systematically: a,e,i,o,u; pa,pe,pi,po,pu; ba,be,bi,bo,bu; etc.).[7]

6 The emergence of writing systems around 5000 years ago started with hieroglyphic/ pictographic symbols and graduated (maybe under demotic pressure) via ideographic symbols to the use of symbols in syllabic reference (to sounds) and eventually to the development of "consonantaries" for expressing the consonants in the North Semitic Language region. This was followed by the emergence of the first complete alphabet, developed by the Greeks but derived from the North Semitic consonantary. The story is complex and some of the details can be found in Diringer (1968) and Fox (2013), (see also Cline, 2014). The marginal relevance here is that from a "cognitive archaeology" standpoint the development of writing moves through a syllabic phase (because syllables are obvious) to a more analytic phase which covers what we now call phones.

7 Note that the sonority accounts of the two sentences offered in 2(4).8.1.2.2 above are not equally easy to segment; the second sentence is sonorous throughout, with minor variations.

2(4).8.2.2 A constituency approach to syllables takes the view that they are composed of vowels and consonants, with some patterning requirements. Typically a syllable will have a vowel as the nucleus, with one or more consonants before and after it in the sequence (or it might just be an isolated vowel). In a language specific way the permissible consonant(s) in the syllable onset and offset, as they are called, will be restricted. In English, for example, a word like *sprints* is monosyllabic with three consonants in both onset (/s/,/p/,/r/) and offset (/n/,/t/,/s/). But note that some other sequences are not permitted (**spritns*, where the * is used to indicate that the formation is unacceptable). Whilst a sonority approach might capture some of this structural detail it cannot capture the tendency in languages to organize syllables to emphasize the onset phase of the syllable. This is reflected as fact in syllabaries, and more generally in the principle known as the Maximal Onset Principle which requires that consonants be attributed to onsets unless a sequential restriction is breached. This too, incidentally, is available to untrained introspection (try identifying the syllable boundaries in the word *introspection*). The principle captures a property of speech, not of any one language, and it is perhaps rather surprising that such detail is available in a generalisation (there are, after all, around 6,000 spoken languages and there would have been more a few thousand years ago).

2(4).8.2.3 Attempting to define the syllable leads us back to phones, and a moment's thought suggests we should not be surprised. If a syllabary is organized as suggested earlier, with a pattern such as {a,e,i,o,u; pa,pe,pi,po,pu; ba,be,bi,bo,bu; ta,te,ti,to,tu; da,de,di,do,du; etc.}, then this is only possible if the underlying analytical insight is operating at the phonetic level – the syllables are not being thought of as unanalysable.

2(4).8.3 Discussion of syllables, in the context of phones and phonemes, requires an analytical approach to speech production and at the same time points to a core observation. Speech is comprised of the sequential arrangement of vocal tract configurations which have certain properties. The open-ness and shape of the vocal tract will produce sonorities with characteristics which permit identification of the specification for producing the tract shape in the most general sense. The precise arrangements of muscles and the timings of their contractions will vary from person to person but the effect is to produce the "right" shape at the "right" time. Likewise, for consonants the configurations are concerned to interrupt the flow of air from the lungs in an audible manner.

The gappiness discussed there aligns with sonority troughs at the base of which may actually be moments of silence.

The analytic stance leads to a view of speech which characterises the manner in which vocalisations are produced, and the place(s) in the vocal tract where obstructions are created, along with something about the manner and timing of the obstructions.

2(4).8.3.1 The characterisations of the vocal tract configurations are nowadays couched in terms of specific/ distinctive features which serve to distinguish sounds from one another. A systematic arrangement of the features permits analysis of all possible speech sounds, and of the systematic sub-sets used in any one language. Furthermore, this analytical approach offers benefits in respect of understanding patterns of articulation found in languages. We turn in the next chapter to consider in much more detail just how these features work to reveal so much, but also how they obscure some of the interesting detail in speech activity. This leads us to think about depicting the organization of vocal tract configurations in continuous speech. One scheme in particular is singled out for generalisation to other aspects of language structure, and thereby eventually for generalisation as an account of sequencing in behaviour.

2(4).8.3.2 It needs to be pointed out here, without going into great detail, that listeners hear speech sounds more categorically than they might appear to be uttered. Experiments with synthetic speech, going back over many decades, have shown this categorical perception effect by means of continuously varied adjustments to parameters controlling a speech synthesizer, which produce "speech sounds" which are nonetheless heard as either phone 'A' or phone 'B' as determined in the experimental set-up. Consider the discussion of VOT presented above (2(4).8.1.2.5) – an experiment could contrive a range of values for VOT grading between /b/and /p/, for example, but these sounds will be heard as definitely /b/ or /p/ except at the categorical boundary between the two where there will be uncertainties and individual variations.[8] For our purposes we simply need to recognize categorical perception as one way around individual variations in speech production. Note, for the child learning the distinction between [p] and [pʰ] the systematic variations need to be learned (for perception and production) and the individual variations need to

8 For those who like this sort of material a more challenging puzzle is offered by Studdert-Kennedy (1982) who examines the dichotic presentation of different synthesized components of a phone and shows that for a given set of stimuli the experimenter can contrive both categorical perception of speech sounds and continuous acoustic discrimination of differences in the same material at the same time.

be discounted by categorical perception.[9] She must learn the two allophones and their contexts of use, but will not be aware of the two allophones and they won't matter for intelligibility or conveyance of meaning. When we discuss binary features, in the next chapter, we will be assuming a categorical account of phones and their differences, not an acoustic account.

2(4).9 Thus far we have focussed in this chapter on speech activity and structures. We need to keep in mind for later discussion that of course spoken language is more than just the spoken sounds organized syllabically. Spoken languages build on the resources offered by the organized production of speech sounds, in sequence, perhaps with rhythm and stress constrained in language specific ways. These foundations are used to build morphological structures of one sort or another, grammatical structures, and pragmatic or discourse structures. There are structures at many different scales of analysis, and these vary from language to language. Once again we need here to finesse the need for a library shelf full of books on morphology, syntax, and general linguistics. Comrie (1989) is a good guide to the details, but for now let us note a few points, extending the notion that there are some language universal tendencies (as in the case of syllables) which help to illuminate the general problem – how do languages exploit the otherwise meaningless patternings of sound? We will return in the next chapter to the organization of those patternings, as the basis for our understanding and modelling of how the Sequential Imperative is served.

2(4).9.1 The first point to note is that the language specific organization of the production of speech sounds – which sounds are used, the deployment or not of allophones, the reliance or not on syllables to indicate stress and rhythm, the tolerance or not of consonant clusters and which, if any, are permitted in the syllable onset, and other aspects not mentioned here – is arbitrary and meaningless as well as being language specific. A child has to learn the French way of speaking, as much as the Dutch (this is not about accent, not mentioned thus far but also a meaningless variation between languages). The repertoire of sounds and the organizational schemes for each language all matter – and children learning more than one language have to master all this. And then, this is crucial, children have to learn that these meaningless sounds are deployed systematically (but arbitrarily) to convey units of meaning, and then that these

9 The distinctions to be learned are the allophonic contrasts – for the language being spoken, but in the context of the differences between speakers.

units of meaning are themselves systematically configured and arranged to convey (more or less arbitrarily) other meanings. By arbitrary here is meant simply that there is nothing doggy about (/d//o//g/), or inevitable about the arrangement of subject before verb before object in a sentence. It should also be noted that the child's learning of the sound/ speech system can proceed through successive re-analysis at different scales to reveal structures and patterns initially learnt holistically (see Peters, 1983; see also 1(3).7.1.1).

2(4).9.2 Analysis and description of units of meaning (morphemes) is termed morphology. There is a simple-minded idea of morphology in a language which has it that morphemes are arranged in sequence. Part of knowing a language is knowing how to arrange the morphemes in sequence to convey the intended overall meaning. In the sentence:

> He slowly gave to her the big blue ball.

we understand that the actor is male and the recipient female; the action was slow; the action is plausibly a passing not a gifting (without further contextual detail); the ball was both big and blue in contrast with a smaller blue ball, and so on. Altering the sequence or part thereof might make it sound quaint, weird, or just plain wrong. A "blue big ball" might be taken to imply the presence (for selection) of other differently coloured big balls, as well as a blue small ball. We can interpret sequencings quite easily. However, not all languages offer such isolated morphemes.

2(4).9.2.1 Semitic languages are unusual in that the morphology is divided into that specified by consonantal roots – for example, triliteral roots such as 'ktb' which pertains to writing and books etc. – and that specified by the patterns of vowels and other devices (vowel or consonant duplication, suffixation). The precise meaning of an uttered word will depend on the template or pattern into which the root is inserted. McCarthy (1981) in an influential paper argued that the morphology in Semitic languages is non-concatenative,[10] with the triliteral roots (as they mostly are) accorded a specific morphological rôle.

Thus we have for the triliteral root "ktb" a set of words such as:

---

10 Interleaved, rather than sequential: an illustration from English can be found when you consider the words *man – men,* and *woman – women.* The plural is formed by changing the vowel, not by appending a sound. A written syllabary would misconstrue the structure in semitic languages. The reader might like to consider that this exercise in "cognitive archaeology" reflects the idea that the alphabet emerged through refinement of an

kataba : he wrote
kattaba : he caused to write
kitaabun : book

and many more – but the point is clear. Morphology does not have to be chained together in sequence (concatenation). In the Semitic case the morphemes are not capable of being uttered as spoken entities – both the roots and the templates.

2(4).9.2.2 Non-concatenative morphology is found quite widely in languages. Systems of agreement (gender, number, case) can be thought of as non-concatenative markers of the gender, number, or case involved. Classification systems can also be viewed in this way, and this fits with the general scheme of multiple specification of output units using a three dimensional perspective on the linguistic structures, which we will encounter in more detail later (see also Aikhenvald, 2003; Craig, 1983; Edmondson, 1989; Lakoff, 1987).

2(4).9.2.3 Another way in which morphemes are not isolated units (capable of independent utterance as a morpheme), even though they are added to words concatenatively, can be found in English. Both the plural formation ("add an 's' ") and the possessive ("add an apostrophe followed by an 's' ") simply add a phoneme (the details will be explored in the next chapter). But the phoneme is not a speech unit (it is not a syllable). The possessive provides an interesting illustration. Consider a sentence such as this (and many similar sentences can be constructed):

The house on the corner's garage burnt down.

The sound is added to the end of a word, in the normal manner, and involves no productive complications. However, the meaning is more complicated, as can be revealed with some bracketing:

[The house on the corner]'s garage burnt down.

The morphological work being done is not quite so simple as adding a sound to a word because it is the phrase (in [ ]) which is involved and which has to be processed to recover where the garage was located (approximately) before it

---

attempt at distributed cognition, which is what writing constitutes, in effect. See FN6 in this Chapter.

burnt down. The phrase boundary has no phonological rôle so although sound is added to the phrase boundary it is the co-occurring word boundary that matters and, as in the case of the plural phoneme, the actual phone produced is conditioned by the word final consonant (we will explore this in the next chapter).

2(4).9.2.4 Comrie (1989) discusses two approaches to analysing languages from a typological perspective. One is a morphological approach – does a language have morphemes which are isolated and discrete, or morphemes which are fused, or morphemes which glue together, or what? Comrie (op. cit.) discusses the details in terms of the index of synthesis and the index of fusion. In relation to the former the two ends of the scale, so to speak, are analytic/ isolating on the one hand, and fully synthetic (one sentence per word) on the other. Comrie (op. cit.) offers as an example Vietnamese with approximately one morpheme per word (analytic). Sign language linguists suggest that a single sign might encapsulate a single sentence (as it would for example in British Sign Language when signing "I slowly gave the big heavy box to her" – although the referent for 'her' and details of the occasion would have to have been established previously, just as in English). The index of fusion sets agglutination of morphemes at one end of the scale, with fused morphology at the other (the examples might be Turkish on the one hand, and Russian on the other). Comrie discusses the interdependence of these two scales and notes that the effort at systematization of language type based on morphology raises many questions.

2(4).9.3 In the various examples offered above we see that the simple-minded idea that units of meaning are strung together in sequence to yield larger units of meaning doesn't really fit the data. Comrie's discussion makes this quite clear. But what of the alternative discussed by Comrie – that of structural patterns as the basis for categorization of languages? Here the focus is on arrangement of words – word-order typology. Typically, the discussion is in terms of the order of Subject, Verb and Object (SVO, for example, or SOV, or any of the remaining 4 options). It turns out that most languages put Subject before Object, and most of those put Subject before Verb. Thus SVO and SOV are widely attested, with VSO being not that rare. The remaining three possibilities are uncommon. There is thus a natural tendency in spoken languages (just as for syllable structure) but it is not exceptionless. Other attempts to look at word ordering as a way of studying language structures and groupings also come up with tendencies rather than absolute universals. More widely – from a linguistic perspective – attempts to find typological conceptions and patterns which are both insightful and convincing are not straightforward (see Comrie, 1989; Croft, 1990; Greenberg, 1966; Hawkins, 1988). It is possible to hint at the difficulties with a few illustrations.

2(4).9.3.1 English offers an interesting device which makes it possible to utter two independent propositions simultaneously (maybe more than two in some circumstances – three being acceptable but more than that becoming confusing). Here's how it is done. There is a sentence structure called a "respectively sentence" which goes like this:

> John and Jill [] are [] 19 and 20 (years old) [].
> Mike, Sue and Anna [] are [] (studying) at Cambridge, Oxford and Durham [].

The [] indicates the possible location for the word "respectively" (which is used only once in a sentence, if at all). The material in () is optional. The syntax of the sentence demands a subject and an object, and in the first example above the nominal phrases "John and Jill" and "19 and 20" fit the bill. The plural form of the verb is required because the subject is plural. Note there is no requirement to add the 19 to the 20 to yield 39. What we actually express in the first sentence are the semantics of two separate propositions "John is 19" and "Jill is 20" fused in an utterance by a syntactic device which delivers a single sentence (the word "respectively" is not required and is commonly omitted). In a case with three or more people mentioned in the subject phrase the structure sounds like a list, and this list is mentally aligned with the listing in the object phrase, to give a stack of propositions all uttered at once and anchored via the verb and the simple sentence structure. This complies with SVO expectations – but the way such sentences work is so different from the simple sentences ("John is 19" and etc.) that typological conformity doesn't provide insights into the conveyance of the meaning of the propositions. The mapping between propositions and sentence is not clear cut, so the value of identifying SVO as the sentence type is somehow beside the point.

2(4).9.3.2 Comrie offers the example of Serbo-Croatian which has free word order. The sentence *Peter reads (the) book today* can be rendered as *Petar čita knjigu danas.* It can also be rendered in any of the other 23 orderings. Comrie notes that including the first person singular dative pronoun *mi* ("to me") normally requires its insertion after the first word in the sentence irrespective of whether or not this is a "sensible" place to insert the pronoun. The problem is illustrated if *Petar* is replaced by *taj pesnik* ("that poet") where the pronoun would follow the first word to give a strange construction ("that to me poet…").

It could also follow the first constituent ("that poet"), which seems more conventional.[11] But that such freedom is permitted, and note that "first word in a string of words" is not a linguistic concept, means that typological clarity is lost – exceptionlessness is not to be had.

2(4).9.4 Taking a larger view over the discussion of language – from the allophones of /p/ through to morphology, sentence structure and word order typology – we see that set alongside the apparently self-evident notion of units arranged in sequence to form larger units, and so forth up to sentences and beyond, is the reality that much of the structuring happens in parallel, and with overlapping entities and structures involving units of different scale. This is an important insight that Cognitive Scientists tend not to appreciate.

It is also the case that trying to find uncomplicated universal statements about language structures doesn't really work (the literature on linguistic typology makes that abundantly clear). We will find a more convincing, although rather restricted, example in the next chapter.

It is perhaps helpful to point out that the ideas on offer here are not closely linked to notions of Universal Grammar (UG). MacNeilage (2008) (as we will discover in Part 3) takes a rather confrontational approach to UG; my preference is to set aside the underlying philosophy of UG and instead just use some core ideas (which were in any case being considered prior to the development of UG – such as segments, features, syllables, and the notion of suprasegmental structures in speech). Notice, however, that the child learning one or more languages has to learn all these details, as well as which matter for what sort of reason – speech production details, alongside morphology and syntax and pragmatics, as well as speaker differences, dialect variations, idiosyncratic production differences, and etc., all these have to be learned and deployed.

We now turn in Chapter 5 to consider the issue of parallel structures explicitly, but also with a concern for the wider perspective demanded by the need to address the way in which we can account for how the Sequential Imperative is served by the brain. This in turn reflects back on language behaviour to show that it is an exemplar for the organization of behaviour more generally. We start with the organization of speech – again – but now look at how the generalities discussed above can be formalised.

2(4).9.5 The material just covered illustrates something of the complexity of speech and language, and in doing so shows the extent of the "dizzying array"

11 The position in a sentence following the first constituent is known as Wackernagel's position.

of factors that have to be covered by Cognitive Science. There are some details to be kept in mind for later chapters. The wide range (continuous) in VOT, for example, amongst the world's languages suggests that simple categorization of entities and identification of segments might not be so simple in reality (despite categorical perception). Likewise, the distributional universals discussed by Comrie (1989) and others are not arbitrarily patterned – some Cognitive Science explanation will be required eventually for the asymmetry discussed earlier (2(4).9.3), but is not offered here (however, see 3(7).13.1.1). The next chapter attempts to bring some coherence to the diversity of data offered, in the sense that multiplicity of accounts can be set aside in favour of some sort of unified process which actually delivers sequencing and de-sequencing. However, it remains the case that a further bigger project could be undertaken which examined whether or not the patterns in distributional universals had some underlying cause in the management of the Sequential Imperative.

2(4).9.5.1 The core idea in the next chapter – with even more linguistic detail – is that the structural detail can be captured in a three dimensional scheme like that sketched earlier in 1(2).5.3.1 (and see 2(4).9.2.2). The purpose of providing the details given above, and the further details in Chapter 5, is not to overwhelm the reader but rather to point to the diversity of linguistic phenomena to be covered in any account. This is sometimes not appreciated by Cognitive Scientists. The possibility of developing a unifying formalism of some sort, derived from linguistic detail but generalized to cover behaviour in all animals, seems so attractive that persevering with the detail must be worthwhile. Of course, the dizzying array of details for behaviour in other animals does need to be documented – there will be complexity in both nature and quantity of data, but we don't have that yet.

2(4).9.5.2 An emerging theme is given more attention in the coming chapters. The syllable, and its rôle in language activity and structure, turns out to be increasingly valued. It could be tempting to elevate the syllable to the sort of prime importance in Linguistics currently accorded to syntax, but it is clear that in fact the interesting thing about syllables is precisely that they assist one to see that language structures are less hierarchical than often supposed. It is also significant that we will see yet again in the discussion of syllables that there are many different ways of envisaging their structures. The variation in componentialities is instructive.

2(4).9.5.3 One point cannot be overstressed; the range and extent of material to be learnt by the child seems to be astonishing. Those who incline to "rule-based" accounts (which supposedly make the learning easier) will find much to ponder because a lot of the material learnt by children achieving language is simply not amenable to a rule-based account. Nonetheless, the appeal of finding concise rules for aspects of linguistic behaviour is very strong, and an illustration is offered which, it is hoped, can be readily appreciated by those who have never encountered such things. The astonishment, it must be said, reflects our lack of integrated conceptions of learning, language, and behaviour. By the end of this book the astonishment should have diminished. The hope is that equivalently astonishing material will be discovered and documented in other species (e.g. dolphins) as well as other primates.

2(4).9.5.4 The reader may be bewildered by the shift from discussion of General Cognitive Principles in the first Part to discussion of phonetic details in speech in the second Part. This is going to get worse before it gets better, unless the reader recalls the explicit mention of the need to approach the problem of making the Sequential Imperative comprehensible and comprehensively workable first by looking at the questions top-down – the Cognitive Science overview – and then by looking up from the details of speech. We will do even more of this in the next two chapters, but the third Part brings the two perspectives together. What should already be becoming clear is the quantity and range of activity to be managed when speaking. The SI has to deliver appropriately sequenced and realised components ranging from durations of a few milliseconds perhaps (control of VOT) through to complex sentences. Every entity or component has to be "delivered" or "assembled" on time and in the right place relative to other entities. This is both Cognitive Science and Phonetics/Linguistics.

CHAPTER 5

Non-linear Phonology and Beyond

2(5).10 Our focus in this chapter is on the notion of speech segment (see 1(1).4.6) and the assembly of sequences of segments. In the 1970s linguists working on the organization of speech reworked some ideas from an earlier generation of linguists, notably Firth (1957), concerning the fact that a purely segmental account of speech is incomplete (see Anderson, 1985). The work of those early linguists focussed on the need for a suprasegmental account, in addition to, if not supplanting, a segmental account. The problem is easy to explain, and this is done below. Arising out of this work a formalism was developed, in the 1970s and 1980s (see Goldsmith, 1976, 1990, 1995), and we will explore aspects of this with a view to showing that it can be adapted to provide the model we need for depicting how the Sequential Imperative can be delivered by the brain. It is a functional model, not a neurological model, and it has great conceptual power.

2(5).10.1 To appreciate the inadequacies of a segmental approach to speech we need first to have a fuller account of the notion of *segment*. Briefly (again the reader is referred to textbooks on the topic, for example Ladefoged and Johnson, 2011; Hayes, 2009) the core concept is that speech is a sequence of speech gestures (vocal tract configurations) which are specified in terms of the articulatory features deployed, segment by segment. The phonologist and the phonetician have to keep in mind two conceptions of what is going on. First, there is a sequence of linguistically specified gestures, called segments or *phonemes* but just as readily understood as idealized configurational targets for the vocal tract. Intra-segmental details and allophonic patterning can be left to the phonetic scale of account, where segments can be considered as *phones* – specific moments in a continuous flow of muscular activity. The vocal tract always has some sort of configuration, of course, but this only matters at particular moments when the configuration is specified as a target, rather than merely being the route from one target to the next (the targets can be dynamic, as in glides). A pedant might want to argue that the vocal tract really only exists as a *vocal* tract when the specified target configurations are achieved. Transitions between targets are for the most part meaningless and thus, perhaps, not really vocal or part of speech – except when they are, so the generalisation is just that, not a statement of a universal characteristic. The idealization is useful here, but doesn't really work, as we will see clearly when we consider

© KONINKLIJKE BRILL NV, LEIDEN, 2017 | DOI 10.1163/9789004342996_009

in more detail the nature of the target specifications and transitions (and see 2(5).10.1.4.2 below).

2(5).10.1.1 The core idea underpinning formal accounts of the production of speech is that the process concerns specification of the manner in which speech sounds are made, and the place in the tract where there is a constriction (full, as in a "stop", or partial). There are some subtleties, but for now let's work with this simple idea because it readily captures the notion that the speech gestures in sequence are concerned to generate a sequence of sounds in the vocal tract – excitations and resonances and their combinations.

2(5).10.1.2 Phoneticians and phonologists have refined their ideas about phonetic *distinctive features* over recent decades – the precise details of the list of places for vocal tract obstruction, and the details of the nature of the obstruction and of the sound source used. Text-books usually come complete with diagrams showing a sagittal cross-section drawing of the head with the lips to the left, say, and the larynx to the right, with places of articulation carefully labelled (cf. Figure 1.4 of Hayes, 2009). Charts are produced showing which features are used to describe which sounds. The idea of *binary distinctive features* originated in the 1950s with the recognition (from information theory) that specifications could be trimmed back to the minimal details required to specify unambiguously any particular speech sound (see Cherry et al, 1953). Say a scheme has 15 binary distinctive features, such as: [±voiced], [±nasal], [±strident], [±alveolar], [±labial], [±lip-rounding], and etc., covering both location and nature of the constriction in the vocal tract, and the sound source(s) involved. We assume the "±" notation to mean that the feature is, or is not, required for the production of a specific sound – "present" or "absent", if you will (the binary aspect derived from information theory). Such minimal descriptions are insightful because they reveal patterns and permit simple accounts of some processes. In particular, co-articulatory effects and assimilations seem rather cleanly described. For example, the fact that the vowel in the word *man* is nasalized because of the nasal consonants preceding and following is rather simply captured, as we will see shortly. However, the simplicity of the featural specification scheme is misleading because when dealing with something like nasal assimilation in *man* (where the vowel becomes nasal because it is surrounded by nasal consonants) the formal account appears to imply that there are three adjacent segments (articulatory targets, remember) which each end up having the feature [+nasal] (along with their other features). This account is mute on the transitions between targets, yet the insight about the assimilation is that nasality is present throughout the production of the whole syllable (word) – nasality has spread to the transitions as well as to the target configuration for the vowel.

2(5).10.1.3 Interestingly, some segments are explicitly concerned with transitions between targets – as when the transition in a glide or diphthong is specified. But other aspects of the temporal structure of the articulation are ignored – so it remains unclear in the formal accounts whether or not a long consonant derives its length from the duration of the stable configuration or whether in fact what is meant is the separation of specifications for the closure and the release (say, in a "stop"). These are two different ways of describing the same thing, but the difference matters in relation to the overall descriptive scheme. If the scheme deployed refers to closure and release explicitly, then other phenomena fit more readily into the account. For example, in colloquial English consonant releases are not always evidenced – this gives the colloquial sound to *red'n'white* where neither the [d] or the [t] is released/ sounded (see also 2(5).10.1.4.2). Phonologists are comfortable with the need to choose descriptive terms or elements to suit the exposition they wish to make, but the result is a sense of variation in emphasis given to different aspects of the speech production process.

2(5).10.1.4 Various descriptions of syllable structure reveal more or less about configurational targets and the arrangement of segments (as we saw above in 2(4).8.2.2.). It is helpful to consider briefly some of these schemes so that the reader can appreciate the lack of certainty in our understanding of something apparently so simple – there is more to the syllable than might at first appear. Earlier we saw that syllables can be defined in terms of sonority contours, and also in terms of constituents. We encountered the Maximal Onset Principal. But how do these ideas mesh with discussions of vocal tract configurations and the conception of speech as a succession of more or less thoroughly realised configurations evidenced by the stream of sound?

2(5).10.1.4.1 There are several structural approaches to the syllable – and the first three we will look at (briefly) are discussed in Carstairs-McCarthy (1999). Syllables can be thought of as having three constituents – onset, nucleus, and coda. Simplistically we can think of these as being the one or more consonants making up the onset, the vowel sound as the nucleus, and remaining consonants as the coda (sometimes called the offset, as we saw earlier). Indeed, a basic account of syllable structure is just this simple set of constituents. Other structural accounts go further and propose that the nucleus and coda together constitute an element called the rhyme; or that the onset and nucleus constitute a single unit of weight with the coda (where present) adding an additional unit of weight (the units are each called mora, with morae as the plural). A light syllable has no coda (second mora) whereas a heavy syllable is bi-moraic. These structural accounts play various rôles in various theoretical accounts of speech, stress, and language more generally. But note that they do not deal explicitly with vocal tract configuration – the accounts of syllables are in terms of shallow "layers" of structure and constituency, ignoring the details of what can or cannot make up any of the constituents. Constraints on sequences of sounds (captured to some degree in the sonority model) are not expressed, nor are language specific phonotactic constraints (specifying permissible sequences of sounds).

2(5).10.1.4.2 In an alternative model for syllable structure, (see Edmondson and Zhang, 2002), I have argued for an articulatory model of the syllable. The syllable structure illustrated below in Figure 5.1 comprises specified configurational targets and unspecified transitions (and avoids talk of segments). The configurational targets may indeed be transitions, but specified, not left to articulatory interpolation. In the diagram we see a sort of recursive expansion of specificational detail, covering ever more fine detail until the point is reached (*tr* transitions) where the specifications are not supplied and articulatory interpolation deals with the matter of getting from one target to the next. It should be noted that in any syllable not all targets have to be supplied – the structure shown is that of a maximally "full" syllable conventionally symbolized CCCVCCC. In this sort of account it is not difficult to see how the specification of some features might be carried over from one target to the next. The feature [+nasal], for example, might be supplied for one of the targets and that value might then be inherited or perhaps passed up the structure to those parts of the structure which do not have specified values for nasality.

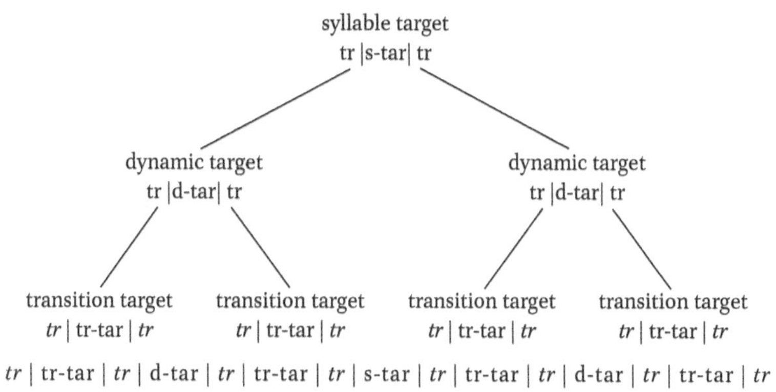

FIGURE 5.1 *This diagram illustrates the structural potential of syllables in a manner which does not rely upon a conventional segmental account, as discussed in the text.*

The structure illustrated can account for some problematic analyses in conventional segmental accounts. The English word *apt* for example sounds clearly as if the initial stop for the /p/ is realised, and the release of the stop for the /t/ is realised, but the transition between the two configurations (a required or specified, "dynamic" target – d-tar – from [p] to [t]) is not sounded. This can be notated as follows:

tr	**s-tar**	*tr*	**tr-tar**	*tr*	d-tar	*tr*	**tr-tar**	*tr*
	[æ]		[>p]		[pt]		[t<]	

where the targets identified in bold are sounded and the d-tar is specified but not sounded. Note that this account is not conventionally segmental although distinctive features could be used to specify each target. This unconventional approach is not used in the literature, but is offered here because it illustrates the need for consideration of sub-segmental detail.

2(5).10.1.5 The above very brief discussion of syllable structure serves to show that whilst accounts of configurational targets can look neat and tidy, the reality is that articulation does not proceed as a succession of twitches or jerks from one configuration to the next. The process of articulating speech is a continuous one on to which is laid an abstraction, usually (but not necessarily) a sequence of segments – and these are what we say are heard (categorical perception) as components of words and utterances (see 1(1).4.6, 1(1).4.9). When the scale of analysis is at the segmental level the segment dissolves into smaller components and eventually into continuity of activity. When the scale of the analysis is broader (phonological) then the segment is a useful but not always obvious approximation and can serve, with the syllable, as the basis for further larger scale accounts. However, our aim here is merely to show that segments shouldn't be taken for granted as the obvious unit of speech on which others are built. Other stories can be told (sonority accounts can be refined) and, as we have seen in the previous chapter, it can make more sense to think of the syllable as the basic unit or element in speech (see 2(4).8.2.1).

2(5).10.2 In order to see how the feature specification of segments works, and how it plays out suprasegmentally, segmentally, and within or between segments, we need only to consider a few features – full feature lists are not required to understand the points being made. There are some details which need some explanation.

2(5).10.2.1 The first detail is that different scholars use different lists of distinctive features (and over time the listings in textbooks change also). The reasons for the differences can be quite abstruse but in essence the effort put into the refinement of a list of features reflects understanding of patterns in the languages being considered as much as understanding of the speech production processes. For example, spoken English does not involve the use of "clicks" which might be found in other languages. There is no point therefore devising a list of distinctive features, for discussion of English, which can cater for such sounds; language specific (or language group specific) listing of distinctive features will suffice to capture phonetic and phonological detail and argumentation.

2(5).10.2.2 The second detail concerns the way in which the distinctive feature characterisations are used. We need to think here about why anyone would bother to note down the absent features, or those not relevant to the argumentation. In some ways the distinctive characteristics are those present, not absent. However, such a radical simplification doesn't work all the time. The nasalization in *man* we discussed earlier is a good example. This can be handled formally with some rather obvious notational formalism. A nasalization rule can be formulated which expresses the insight that the vowel in *man* is nasalized because the consonants either side are nasalized. This is illustrated below (Figure 5.2).

$$\begin{bmatrix} -\text{nasal} \\ +\text{voice} \end{bmatrix} \longrightarrow \begin{bmatrix} +\text{nasal} \\ +\text{voice} \end{bmatrix} \Big/ \begin{bmatrix} +\text{nasal} \\ +\text{voice} \end{bmatrix} \underline{\hspace{3cm}} \begin{bmatrix} +\text{nasal} \\ +\text{voice} \end{bmatrix}$$

FIGURE 5.2 *Schematic representation of a rule for nasal assimilation.*

The rule states that an un-nasalized voiced segment (the bit on the left in extended square brackets) becomes (the arrow) nasalized, in the context (the bit after the big forward slash) of having a nasalized voiced segment both before and after it (the line between the two segments). [The use of a rule to express the account of the data is for concision, not because it is supposed that producing speech requires rules to be followed by some sort of mechanism.] The specifications of the segments in the rule as shown do not distinguish between vowels and consonants – that could be done by adding more distinctive features. But the rule does illustrate one thing quite well – it would cover the words *mam, mom, nan, men, nam,* etc., just as well as *man*. In this sense it shows clearly the power of such rules in capturing generalisations and processes. And it shows what is and is not achieved by leaving out some of the features, both positive and negative values; greater generality with fewer positive features, greater clarity with the negative features left out, except of course where a change in value is the point of the rule. And, note the rule says nothing about the nasality of the transitions.

2(5).10.2.3 A rather extreme, but attractive, example of how concise one can be when expressing phonological details in rules is the following. Apart from exceptions controlled in other ways the adjustment required to make an English noun plural is to "add an *s* sound to the end of the word". We all know that. The exceptions might be words like *mouse, ox, child, woman,* and so forth. But the more regular words also fall into patterns, if one thinks about the process and does some work in front of a mirror, and with one's hand on the throat. There is a group of words like *cat, pup, stick,* ... which all end in a sound which is not voiced [−voice]. There is another group which end in sounds which are voiced [+voice]: *dog, club, fad,* ... When the words in the first group are made into plural forms the sound added or appended is indeed an *s* sound. For the second group, however, the sound appended turns out to be a "voiced *s*" which we call a *z*. There is yet a third group of words – those which end in hissy sorts of sounds, including *s* and *z* but also the sounds at the end of words like *church, judge, dish, garage*... In this group of words each word ends with of a sound rather like the plural sound to be added – and this causes a problem of discriminability (*busssssss* still sounds like just the one bus, regardless of how much noise you make at the end of the word). The solution is to add a vowel sound (a syllable nucleus) to which can be appended an appropriate sound – *z* because the vowel is [+voice]. This gives us *busiz* (spelt "buses") although in some dialects a bus is a *buz* and the plural of bus is *buziz,* which is indistinguishable from "buzzes".

2(5).10.2.3.1 Let us look at what is going on here in more detail (!). It is possible, using distinctive features, to capture the totality of the above description in a relatively simple phonological rule. Figure 5.3 covers the "add an *s*" notion of making plural forms of unexceptional nouns in English. The first component of the rule – i) – states that if the word ends in a sibilant sound (the [+sibilant]

$$\text{i)}\quad \emptyset \longrightarrow \begin{bmatrix}\text{+syllabic}\end{bmatrix} \Big/ \begin{bmatrix}\text{+sibilant}\end{bmatrix} \quad \#\_\_\_$$

$$\text{ii)}\quad \begin{bmatrix}\text{+alveolar}\\ \text{+sibilant}\end{bmatrix} \longrightarrow \begin{bmatrix}\pm\text{voice}\end{bmatrix} \Big/ \begin{bmatrix}\pm\text{voice}\end{bmatrix} \quad \#\_\_\_$$

FIGURE 5.3 *This two-part rule accounts for the "add an* s*" data in English.*

feature – followed by the # symbol to indicate the word boundary) then immediately after that sound place a new syllabic nucleus: this is expressed rather quaintly as "nothing (the ø) becomes (the arrow) a vowel [+syllabic]".

In reality the inserted syllabic sound will be a sort of general neutral vowel sound – sort of like *ih* or the vowel sound in the word "kick" (controlled by the high position of the tongue for the sibilant sounds either side of the vowel – this is another example of assimilation; but we don't need to show all the distinctive features which specify the vowel or its predictable assimilation to the sibilant sound tongue position). The second component of the rule says that a specific sort of sibilant sound, made with the vocal tract constriction at the alveolar ridge, just behind the top teeth (*s* or *z* – not fully specified on the left of the arrow) is added to the end of the word (marked #). [Part i) may not have been triggered but if it has been applied the word will end with the extra vowel sound spelt "e" and sounding like *ih*.] Voicing will match the voicing at the end of the word (this is what the duplicated [±voice] means).

2(5).10.2.3.2 Four things are worth noting about the rule just introduced, apart from its stark simplicity and elegance. Firstly, the rule is a description, not an instruction. There is no suggestion that we "run the rule" on some sort of phonological calculator every time we generate a plural – it merely looks that way. We know the required patterns – we know what sounds can and cannot follow other sounds. We don't know we know all this – most people are completely unaware they know so much detail about control of the articulatory apparatus. And of course we learn it without knowing we are doing so.

Secondly, the features listed are not universally agreed. Hayes (2009), for example, prefers [+strident] for [+sibilant] but the difference is minimal – both refer to the quality of the sound produced by a not quite closed constriction (the manner of the constriction), which causes turbulence which excites the vocal cavity and produces recognizable speech sounds (as in whispering). In contrast, the feature [+alveolar], which refers to the ridge in the roof of the mouth just behind the top teeth, says where the constriction is to be found (place, rather than manner).

Thirdly, the feature [+syllabic] is slightly mis-used – it refers normally to a consonant which is being used as a syllable nucleus (the *n* in *fish'n'chips* or *cut'n'paste*). However, it makes the point, with the location of the tongue providing the vowel quality derivable from adjacent configurations/ segments. The first part of the rule is about vowel epenthesis (insertion of a sound), as it is called, and the use of the feature [+syllabic] is a simple way to capture this. Another way of formalising the epenthesis would be to specify that the vowel inserted is a particular neutral vowel sound, and use the phonetic symbol itself in place of one or more features. My view is that this is not as insightful as the rule given above.

Fourthly, the rule actually captures a phonotactic constraint which is more widely used than just for the formation of plurals. It turns out that the restriction on sound sequences captured in the rule, and expressive of what happens when we "add an *s*", is also found elsewhere in English when we "add an *s*". So, for example, adding the *s* sound to indicate possessive – as in the example earlier about the *house on the corner's garage* burning down – is also controlled in the same way (and likewise the possessive in *Stan's dog*, introduced in 1(2).5.3.1.1). And the addition of the *s* in third person singular present tense verb form is controlled in the same way (I eat, he eats; I add, she add*z*; I buzz, it buzz*iz*). We know the phonotactic constraint with increasing completeness from young childhood through to primary school (Berko, 1958). However, Hayes (2009:193) notes that there is a plural deviation for words patterning with a final "*f*" becoming voiced as in *loaf – loaves*, *wolf – wolves*, and etc., with *loafs*, *wolfs*, sounding acceptable for the third person verb form (*he loafs around, she wolfs her food*). The complexity of the use of "add an *s*" becomes apparent when we consider a set of sentences like:

> She gardens in the rain.
> The garden's slope makes mowing difficult.
> The gardens open tomorrow to visitors include several in the village.
> The garden's open tomorrow to visitors.

Note that the contracted form of the verb *is* is governed by the same constraint. In all four cases the sound is *gardenz*. We don't recover the meaning of the *z* until perhaps quite late in processing the sentence (consider the last two sentences).

2(5).10.2.3.3 The apparent digression into the intricacies of speech production is merely to show just how much we have to do to assemble sequences of sounds when we speak. As well as the dizzying array of muscles to control, there is the complex patterning of the sound system to be learnt and deployed for linguistic purposes. And all this and learning grammar, and the meanings of words, too!

2(5).10.2.4 One problem with any comprehensive listing of distinctive features for speech segments is what do you do if a feature is irrelevant in the production of a particular sound? A good example in English is lip-rounding as a feature in the production of consonants. Lip-rounding is a feature required for the production of a vowel such as the *oo* sound in English words like *who, true,* and the supposed cow noise *moo* (try it, looking in a mirror; a stiff upper lip will suppress the rounding). It is widely encountered, so the topic is not an esoteric side-issue in the application of distinctive feature analyses to English speech. Some listings of distinctive features show [–lip-rounding] as the feature for consonants (all consonants). Others show the feature as unspecified for consonants, which seems to be more subtle in its effect. The two accounts lead to different sorts of insight and claim in respect of spoken English.

2(5).10.2.4.1 Where the feature [±lip-rounding] is given a negative value for all those sounds for which it is not positive, then a rule is required (in the account, not in the brain) to change the value in some circumstances. The word *construe* in English is a good illustration of the need to think through the options. The second vowel – the *oo* sound – requires lip-rounding. If you pronounce this word whilst looking in the mirror you will notice, in all probability, that the lip-rounding starts in the consonant cluster; probably not for the *n* sound, but certainly in place by the *r* sound. A rule could be contrived to "spread forward" the lip-rounding, segment by segment – until some condition is met. It could be cumbersome and awkward to formulate – a sure sign that an insight is not being captured properly; sure, the notion that lip-rounding is being set-up in anticipation of the vowel is captured, but the way this is described is likely to be clunky.

2(5).10.2.4.2 The alternative is to take a step away from the chain of segments and think about the idea that lip-rounding might be a (suprasegmental) feature of the syllable, one that is supplied by the specification of the syllable nucleus, the vowel. There is then no need to worry about the value of the feature in relation to consonants (a specification is not required). In this model the vowel supplies the feature which is applied to the *syllable*, including the consonants in the syllable onset. Try it again in the mirror – does the lip-rounding appear for or during the *s* sound, to be well in place for the *r* sound and then the vowel? The account has simplicity, but of course the dataset is limited to just one word. However, the point here is not that the syllable domain is the right domain for lip-rounding as a feature (see the discussion in Coleman, 1992) but rather that thinking about features, and rules for spreading/ changing values, does not have to be limited to the notion of the segment. The nasalization in *man*, discussed earlier, tells a similar story – this time with nasalization spreading across the nucleus of the syllable from onset and coda; the account is syllabic, not segmental, and thus covers the transitions between segments automatically.

There is, in addition, another aspect of segmentation highlighted by thinking about features associated with syllables rather than segments. The notion of the segment boundary, and indeed the syllable boundary (and others), has to be nuanced. We do not find, nor should we expect, a precisely synchronised and time-aligned set of value changes (from, say voicing to voiceless, aligned with onset of frication, release of a stop, and etc.). The imprecision in timing of the start of lip-rounding (or more accurately, its achievement of the target configuration of the lips) provides evidence of the fuzziness of the alignments we consider to be simultaneous. This fuzziness is the norm. We will encounter more examples throughout the remainder of this Part.

2(5).10.2.5 We have already noted (2(4).8.1.2.5) that voice onset time is a sub-segmental detail which serves to categorise segments (as aspirated or not). VOT can vary over quite a wide range, and the values to be deployed in any one language have to be learned and controlled for successful use, but the outcome can be category distinctions in perception. The convention is that the segment concerned is aspirated (a distinctive feature) but in fact the aspiration is a feature related to the transition to the voiced segment which follows – it is in essence a property of the release phase of the voiceless consonant concerned which determines the interval before the voice onset of the following vowel (it is control of the fuzziness in timing). We can also note the utility of distinctive features in accounts of segments where a feature value spreads to other segments to create specifications for longer stretches of speech, whether nasalization or lip-rounding in English, or some other phenomena in other languages. In all such cases the speech segment is revealed as an abstraction from the continuous flow of articulatory activity. Speech is a succession of gestural targets in the configuration of the vocal tract – simultaneously the continuous flow of activity and the sequencing of recognizable specifications known as segments.

2(5).10.3 We return then – with a little more insight into how linguists describe the complexities of speech – to the point made at the outset of this chapter: speech is both a sequential arrangement of entities and a continuous activity. What matters is the scale of the description being offered and the rôle of that account in other characterisations. Phonology can be thought of as the broader scale account where the finer detail available to the phonetician is not required – the difference between thinking about /p/ in English, versus [p] and [pʰ] in English. What we turn to now is a formalism developed for phonological theory covering suprasegmental phenomena, but which turns out in principle to be scale independent and thus able to cover all sorts of other phenomena. It is called "non-linear phonology" (or, more technically, autosegmental phonology, see Goldsmith, 1976, 1990, 1995) and is readily described on the basis of the brief exposure to the ideas and formalisms in the preceding paragraphs and chapters. We will set out the material in two ways which, when taken together, reveal the power of the conception as a descriptive device.

2(5).10.3.1 Recall the Naum Gabo style of structural account provided earlier (Figure 2.2). The core idea is that the sequence of segments in the spine – the central string of blobs running diagonally up that image – is labelled, section by section (or word by word, etc.), in a fashion which has the labels all arranged simultaneously around the spine without any one of them preceding the others. This concept is at the core of the formalism developed for non-linear phonology, but the diagrams look very different.

2(5).10.3.1.1 The diagrams have three essential components – the units arranged sequentially along the spine, typically from left to right; an arrangement of half-planes anchored at the spine and radiating outwards (these half-planes are called tiers in many accounts; we will use both terms but the reader should be aware that the word tiers comes with connotations of hierarchy whereas the more geometrical conception – half-planes – does not); values placed at the edges of the half planes with links to the units in the spine (these are the labels). The units in the spine are typically indicated as unspecified segments (marked with an X), or as vowels and consonants (V, C). A diagram which illustrates the spreading of lip-rounding in the word *construe* is offered in the illustration below (Figure 5.4).

Letters of the alphabet are used for the sounds instead of phonetic symbols (see Ladefoged and Johnson, 2011). Note that the representation *konstru* is unconventional, but suffices here. The diagram offers a mix of details to illustrate various points.

2(5).10.3.1.2 The diagram is to be seen as being a representation of a three dimensional arrangement of three planes radiating out from the central spine (shown as XXXCCCV). The first syllable has XXX in the spine instead of CVC.

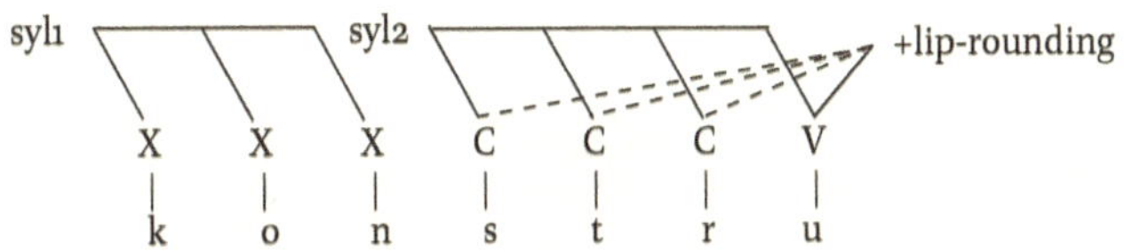

FIGURE 5.4 *Sketch showing three half-planes linked to a string of segments in the spine. The spinal segments are either Xs or Cs and Vs; this is discussed in the text.*

How does the reader know that the segments are CVC? In a more complex diagram there might be two more tiers, one for labelling segments as consonant, the other for vowels (and you could label a segment as both vowel and consonant if you wished – the syllabic "n" discussed earlier for example (2(4).8.2.1)). In a fuller scheme the identification of segments as "k", "o", "n", could also be replaced by a set of half-planes – one for each distinctive feature – and this would cover the distinction between vowels and consonants automatically, but the number of half-planes radiating out from the spine might look overwhelming. In the diagram above only one feature has been shown with its own tier – the feature [±lip-rounding] is shown, with a positive value for "u" spreading forward (dashed lines) to the consonants in the second syllable onset. The lines linking values or labels to segments are shown in two versions. The syllables are indicated in a rather abstract way that emphasises the half-plane geometry; the lip-rounding value is shown in a more conventional "fan" arrangement but if there are many of these the organization of the half-planes can be confusing. Indeed – the presence of too many half-planes in any of these diagrams will be confusing, just as the presence of too many features in a listing can be distracting (cf. the discussion in 2(5).10.2.2 above).

2(5).10.3.1.3 The diagram above has a built-in property which blocks generalisation – the units in the spine are speech segments. There are two ways to think through the issues raised by this observation. The formalism was developed for work on the phonology of speech and thus unsurprisingly incorporates the notion of a sequential arrangement of speech segments. If the core concept is to be extended – even generalised – starting with development of a version to be used for signed languages, say, then perhaps all that is required is to find the equivalent in signing of the spoken language segment. Attempts were made to do this in the 1980s and 1990s (see for example Sandler, 1989, 1993) although alternative conceptions were proposed (Edmondson, 1986b, 1990). The work is complicated by the fact that signing involves both hands, as well as other parts of the body (do you need a spine for each manual articulator?). In many cases signing also involves quite a lot of mouthing, sometimes derived from the visible aspect of speech (lip-patterns especially) which appears to have conventional speech segment characteristics. Work on signed languages serves to highlight the fact that even when used for speech the formalism incorporates some simplifications or idealizations. For example, a long segment may need to be expressed as occupying two segmental slots in the spine (note that the actual duration of a segment in the spine is not specified and can, in principle, vary from segment to segment). Additionally, the formalism is definitely concerned with phonological generalisations rather than phonetic details – there is no way in the formalism as it was developed to capture all the timing subtleties of voice onset timing and aspiration, as discussed earlier (2(4).8.1.2.5).

2(5).10.3.1.4 The alternative way to think of generalisation is to recognise that the speech segment – or the sign, or whatever is being incorporated into the formalism – is actually just a period of time for which there are some specific and systematic characterisations available. The spine then needs to be reconfigured as comprising the smallest units to be found in any behaviour (in any species, ideally). The simplest way to do this is to pick arbitrarily small units of time, for example the millisecond, and say that the elements in the spine are just "millisecond segments of time" arranged in sequence. It may transpire that there are never occasions when behaviour is regularised at that level of temporal refinement – in which case the minimal unit of time in the spine could be made longer; that is an empirical matter. Once the spine is reduced to marking the passage of time, and characterising behaviour and events at that scale of detail, then the formalism becomes general and can indeed incorporate the simultaneous use of speech and gesture, say, or signing, or whatever. In such a scheme speech segments, one by one, need to have their timing details provided alongside the distinctive feature characterisations, alongside phonetic details such as the relative timings of the movements of articulators, voicing and etc. Indeed, should anyone seek to do fine scale analysis of the patterns of muscle activations involved in speech (many muscles, quite fine control of timing) then the formalism would serve that as well as something more coarsely grained such as speech segments. But, and this is important, the entities, segments, activities.... recorded in the spine and which constitute the behavioural segments of interest will always be reducible to the minimal unit if desired. And the behavioural units of interest can be as large/ long as required for accounts of behaviour (as well we see shortly).

2(5).10.3.2 The second characteristic of the three dimensional diagrams which is important, but not so far discussed or made explicit, is the linking between half-planes and the relationship between the non-linear model represented in the diagrams and more conventional hierarchical models (cf. the distinction between the two diagrams provided earlier (Figure 2.1 & Figure 2.2)). A set of characteristics for an entity in a speech stream (the entity could be a conventional speech segment, or a syllable, or a morpheme, or a word...) can be envisaged as simply a list, or set, of features or specifications – supplied in parallel, as it were. Or it can be thought of as the culmination or summary, via inheritance through the nodes of some hierarchical tree-structure, of the relevant specifications or characterisations provided in the hierarchy. The fact is that the use of a hierarchy expresses structural details which could otherwise surface as yet other parts of the specification on yet other half-planes. Statements of relations between (extra) half-planes can *always* replace the incorporation of structures in hierarchies (involving just one half-plane). The *presentational* trade-off is thus richness of account within a "complex" half-plane versus richness of account involving many "simple" half-planes. We will consider some illustrations of this point below.[1]

2(5).10.3.3 The diagram in Figure 5.4, with the half-planes radiating out from the central spine is the structural basis of the generalised system proposed in this Part for accounts of other language behaviour, and indeed other behaviour more generally. However, such diagrams are limited in two ways which are difficult to resolve diagrammatically: they do not provide a process account, although they can illustrate a process (see the earlier account in 2(5).10.3.1.1 showing spreading of lip-rounding in a syllable); they do not depict relationships between half-planes. We can discuss these two issues briefly, and illustrate the problem with a couple of further diagrams.

2(5).10.3.4 It was pointed out above that the radiating half-planes version of the formalism could always capture the relationships between different scales or schemes of description of entities. Tree-structures of whatever kind can be replaced by more half-planes, with statements of the relations between the half-planes. This is true, but difficult to represent diagrammatically. We consider two partial solutions below. The representation of processes by which segments acquire (or lose) values is, as already noted, problematic. The lip-rounding example shows a partial solution but avoids explicit mention of key ideas.

2(5).10.3.4.1 The portrayal of structural complexity in the diagrams can take two forms, both motivated by the push for clarity or readability (at the cost of comprehensive detail). In Figure 5.5 (below) the view of the spine is "end-on"

1 There is one geometric alternative to both the simple and complex half-planes just discussed which can be considered, if only for completeness (see Hayes, 1990). A proposal made during the development of non-linear phonology was that the organization of tiers could itself be hierarchical, so that the relationships between the various features could be represented explicitly in the appearance of the formalism regardless of the values for any speech segment. This produced some odd looking diagrams and embedded speech articulation in the core of the formalism. The view adopted here is that the radiating tiers/ half-planes around the spine – Hayes' alternative model – is the more useful because it is generalizable.

with the half-planes radiating out from the centre. The depiction on the left is of a set of half-planes – perhaps those specifying speech segments – one of which projects beyond the outer-circle in the diagram.

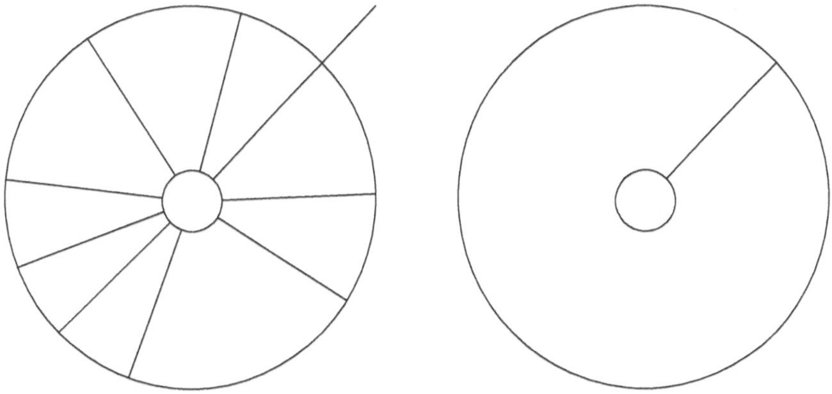

FIGURE 5.5 *Sketch of spine "end-on" showing how one half-plane extends from the spine further than the rest and can become representative of the rest in the sense* primus inter pares.

It could be envisaged that the arrangement and values specifying the segment at any one time are somehow summarised by the value on the half-plane projecting beyond the outer-circle (perhaps just as a V or C), and this half-plane then "represents" the segment in a different diagram, at a different scale, as shown on the right (to which might be added more half-planes concerned, say, with syllable structures or lip-rounding or whatever). Of course, the problem here is that when the diagram is re-orientated to show the spine horizontally the complexity of three-dimensional representation is likely to be overwhelming. However, the diagram does capture one interesting idea, namely that a group of descriptors could be represented elsewhere in the scheme by just one of them – *primus inter pares,* so to speak. This is, in essence, what is done when labelling the segments in the spine with Cs & Vs, instead of Xs (see Figure 5.4).

2(5).10.3.4.2 The notion of representation introduced here is not motivated by the desire to limit clutter in the characterisation of a segment or otherwise to focus just on the features of interest in any account (cf. the discussion in 2(5).10.3.1.2, 2(5).10.2.2). Rather, the idea is that natural units at different scales of description need to be identified in the overall scheme. Thus, we speak of speech comprising a succession of vowels and consonants without having to go into great detail, and of syllables being important units in speech for stress or rhythm assignment without going into great detail. We do need to know sometimes if the syllable is "light" or "heavy" – a complication noted earlier (see 2(5).10.1.4.1) – but nonetheless it makes sense to consider the syllable as a unit the detailed substructure of which is not explored when considering larger units. The duality of cognitive entities (GCP2;(ii); see 1(2).5.2) is our concern here. It is important to note that syllables are the smallest unit which have the property of being entities which have sub-components, and which also serve as sub-components. In this sense syllables are the basic linguistic unit – arbitrary and richly exploited in complex structures. The speech segment is not a basic unit in the same sense – the *phoneme* and the *phone* are linguistic, but they are not systematically deployed cognitive entities with linguistic sub-components, as the articulatory features (and values) are not components in that sense. The example of the "add an *s* " notion for plurals and so forth, discussed earlier (2(5).10.2.3), illustrates morphology leaking down to find expression through smaller entities than the syllable. Indeed, syllables and morphology do not necessarily align in languages but this does not require assumption of the sub-syllabic components as the basic units in language. Speech production (and arguably sign production – see 2(5).10.3.1.3) is behaviour organized syllabically.

2(5).10.3.4.3 The second way in which complexity in the diagrams we are discussing can be reduced to aid clarity of interpretation is to accept a rôle for tree-structure depictions. Consider Figure 5.6 below.

It was noted earlier that any information or linguistic detail captured in tree-structures could as readily be captured in the new diagrammatic scheme through the use of more half-planes, with linking of values between half-planes. And that such half-plane focussed accounts avoid some of the limitations of tree diagrams. So why re-introduce tree-structures back into the scheme? They can be tolerated as a short-hand for the depiction of a complex arrangement of half-planes; they de-clutter the diagrams but can (re-) introduce unwarranted assumptions. We can look at an example of this sort of mixed diagram to see how well it works. In Figure 5.6 we see a respectively sentence (2(4).9.3.1) with a syntactic tree-structure on one half-plane, and two propositions one each per half-plane. The diagram clearly shows the two independent propositions and the single syntactic structure. The alternative involving many half-planes could look difficult to understand.

2(5).10.3.4.4 Some linguists working in syntax have contrived 3D structural diagrams for their theoretical work. Van Valin's work on Rôle and Reference Grammar (RRG), for example makes use of such diagrams. The front cover of

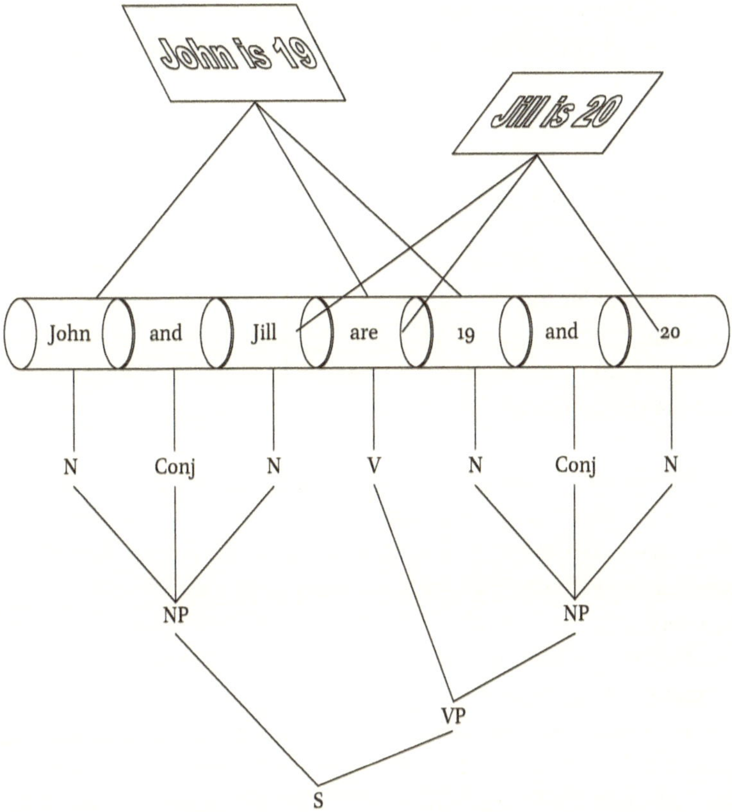

FIGURE 5.6 *Sketch showing different information in three half-planes around the spine.*

Van Valin (1993) has such a diagram, and they are deployed elsewhere in the work. The diagrams in this case are not motivated by work on phonology, they just seem right, in the context of RRG.

2(5).10.3.5 We can summarise the argument in this chapter thus far as follows. The three dimensional diagrams of non-linear phonology can be generalised away from speech specificity to cover other aspects of language behaviour by (a) changing the basic units in the spine to be something short enough to cover all relevant behaviour (for example in control of voice onset times, fine control of all the articulators used in signing, whatever); (b) adding extra half-planes around the spine to deal with relevant detail such as propositional content, pragmatic structures, etc. (these are not necessarily speech specific). A third issue is that (c) the diagrams represent structures (data structures) rather than processes. The latter are captured as combinations of "before" and "after" structures, a simplification which reflects the thinking behind formulation of phonological rules rather than understanding of behaviour. And the fourth issue (d) is that the diagrams become very complex as the richness of detail is increased.

2(5).10.3.6 Alongside these caveats it remains the case that the formalism can be recognized as having generalised potential, if somewhat worrying complexity. The concept looks workable in principle, even if not in any simple paper-based technique. Consider the illustration Figure 5.7 (on the next page) which emphasises the scale independence, and also the magnitude of the representational challenge. (N.B. Time is shown running right to left to encourage the reader to examine the drawing.)

2(5).10.3.7 In this sketch we see scale independence (the drawing could be vastly extended in both vertical directions to cover timing units in the spine ranging from milliseconds to the longest unit thought relevant for the matter in hand – easily minutes, certainly hours, probably days, possibly years). This translates to a very large number of tiers/ half-planes radiating away from the spine. Indeed, the latter point arises in connection with control of the dizzying array of muscles (2(4).8.1.2.7) involved in speech (and likewise sensory inputs). The diagram also illustrates schematically the general notion of sequencing and de-sequencing (bi-directionality within the one "mechanism" – we will consider this later), and the location of the activity within a generalised background of processing (BGP) which is envisaged as the atemporal context for the sequencing and de-sequencing (cf. the comments on the difference between Figure 2.1 and Figure 2.2).

2(5).10.3.8 The diagrammatic illustration of the model (Figure 5.7) introduces an additional element. The BGP are processes, not just repositories (as was suggested earlier – see 1(3).7.6). This becomes important, as we will see, but for now the reader needs to keep in mind that the (de-) sequencing is not the only processing going on. The (de-) sequencing relates to real world timings and sequences; the activity in the background processes is "off-line" and not aligned in any way with the sequentiality considered thus far. This does not undermine the conception of the Sequential Imperative. What is implied by the activity in the BGP is simply that the processing going on there involves the creation or modification of cognitive entities without necessary reference to the organism's activity and engagement with the world. This important point is discussed further in Part 3.

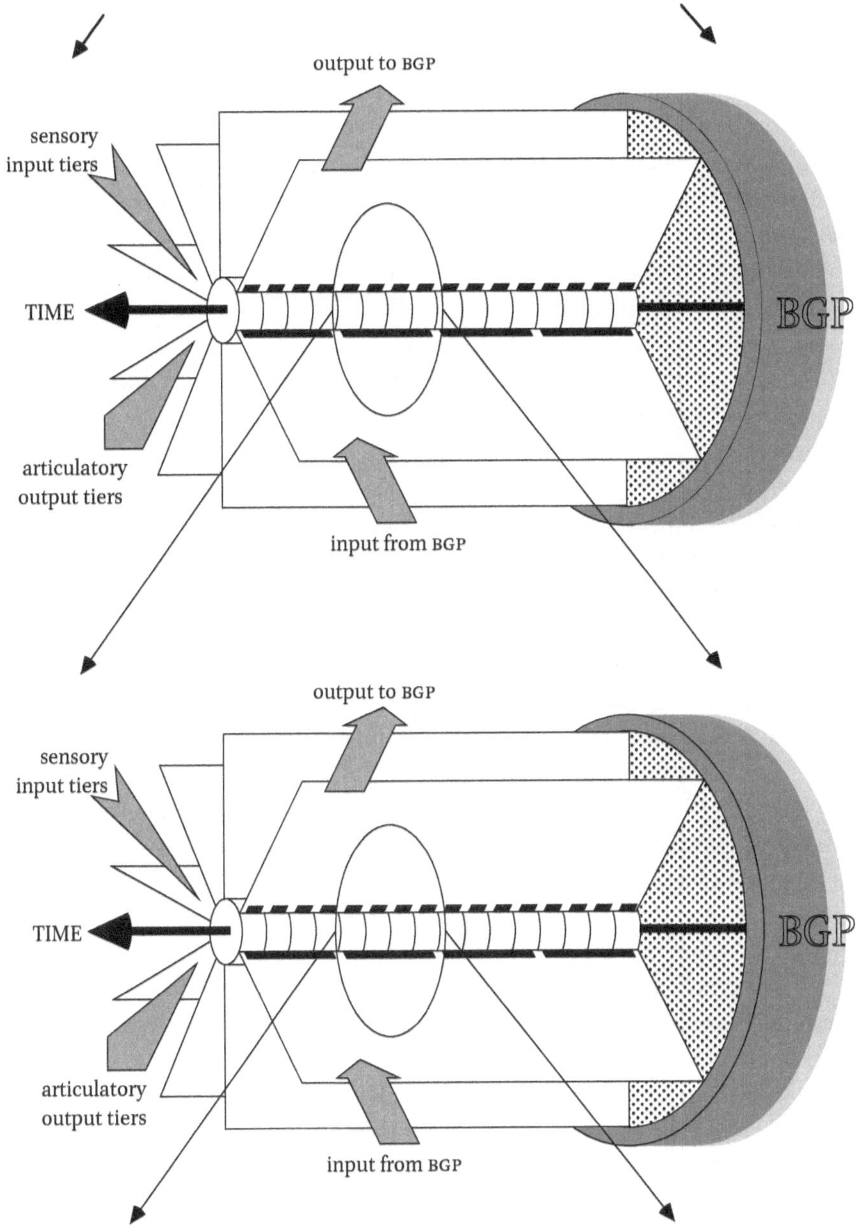

FIGURE 5.7 *This is a sketch of the operation of the model of the (de-) sequencer. The reader is to imagine the scale changes in the spine consequent upon following the diagram vertically beyond the page (in both directions). In addition, the reader should recognize that the units in the spine are only depicted with uniform durations for graphical convenience.*

2(5).10.3.9 The diagram does not show the processes on each half-plane which "read" other half-planes via the data-structure in the spine, and which control insertion of values into the spinal data-structure. The point here is that taken in totality the "formalism" on paper ceases to be tractable (even with some simplifications such as use of tree-structures) when applied to a large range of activities with different scales, or when applied in detail to the physiological complexity of speech production. Fragments can be readily portrayed (cf. the lip-rounding example earlier) in a conventional way, but thoroughness of account, involving many hundreds of half-planes or tiers – indeed a dizzying array of half-planes – to say nothing of the complexity of the processing and the context, is beyond paper and pencil. The idea is clear – the conception is appealing and powerful – the realisation is challenging. We turn to this in the next chapter.

2(5).10.4 The core idea in this Part is the development of a three dimensional model or architecture which can be the basis of an account of behaviour and perception, in all creatures with brains. The model described in this chapter is generalized from non-linear phonology. It can be envisaged, but not readily deployed with pencil and paper. It is easier to see the model as providing some sort of complex data structure than it is to see it as expressing processes of action and perception – this will be covered in remaining chapters.

2(5).10.4.1 The model emerges from consideration of the fine details of speech. It comes from spoken language phonology, and we have seen some of the scale of detail that has to be considered in phonology. However, the model appears readily capable of generalization to other language behaviour – signed language structures, syntax, etc. The leap of faith is to recognize that the model can apply to non-linguistic activity. Perhaps for some readers the leap is readily achieved if you think first of the model capturing and integrating the details of speech production and perception along with all the behaviour sometimes called para-linguistic: gesture, facial expression, eye-gaze, hesitation, as well as more obviously pragmatic aspects of prosody (inflection, mimicry, etc.), and including contextualization. If it can do all that, and deal with speaking whilst chewing gum, walking, and putting up one's umbrella.... then it really is general purpose.

2(5).10.4.2 Perhaps the easiest way to appreciate the domain independence of the model is that it is scale independent. It works the same way with writing a book as it does with uttering the sounds for the word "book". Recognizing a bird song is managed in the same way as tracking an animal through a forest. The bigger the brain, the more half-planes in the system model (we will explore this later) and the larger/ longer the cognitive entities involved; but the principle is the same. The specifications of articulator/ muscle movement, and the "reading in" of sensory data (i.e. the sequencing and de-sequencing) are available in the temporal flow of events, milliseconds, whatever. The co-occurring cognitive precursors to action are the cognitive entities, at all scales, and likewise the sensory de-sequencing feeds into many different entities at different scales (but all co-occurring). The functional specification of the brain does not change with size because the Sequential Imperative is independent of brain size as well as scale independent in respect of actions and perceptions. We suppose that the functional model illustrated above reflects the way the brain serves the SI even though the model is not couched in neurological terms.

2(5).10.4.3 Between the laboured tea making invitation in Chapter 1 to consider the details in behaviour, and the detailed material on speech production in this and the preceding chapter, comes the need to consider the details of behaviour in other animals. But the data are woefully thin on the ground. One can envisage some of the domains which would repay detailed analysis: non-human primate social behaviour, dolphin behaviour (horrendously difficult to study in detail), bird behaviour. We must assume that there is much more to be discovered and that the complexities we see in human activities reflect successful documentation rather than profound differences in kind. For those who remain unconvinced that a model derived from speech can address behaviour in other organisms there nevertheless remains the need for understanding how the Sequential Imperative can be served in all brains. To propose a solution for humans whilst arguing that different solutions are required for other organisms does not seem to make much sense. The suspicion must be that behaviour in other species is indeed complex but just not studied in detail. Tool usage in non-human primates needs really detailed study – there will be dizzying arrays of muscle activity, surely, and the behaviour will be contextualized, and the animal will be attentive to other conspecifics around it, and so forth: complex behaviour. But one must also guard against over-generalization from exceptional cases such as Kanzi's language behaviour. (See Savage-Rumbaugh et al, 1993, for an account of Kanzi's linguistic skills).

2(5).10.4.4 The reader may well be exhausted by the details on offer in this and the preceding chapter. They matter, as explained earlier, because they reveal just how complex behaviour can be, and thus how complex any model must be. The next chapter in this Part deals with the possibility of computational realisation of the model, both as formalism and as speech engine for speech processing. In subsequent chapters the model will be referred to as if it is somehow real in the brain. The sleight of hand involved is not worrisome – it amounts perhaps to overuse of the metaphor, and in any case from time to time I remind the reader explicitly of the issue. The model is best thought of as a way of envisaging the sort of thing the brain has to accomplish, but without worrying about the neurological detail. It is a functional model and as such brings to occasions of its use all the background analysis of details of speech articulation. The details matter because they underpin the model which can be used without the details, but in the knowledge that the details are there should they be needed.

CHAPTER 6

Building a Model

2(6).11 The discussion at the end of the previous chapter drew attention to the size, or extent, if you will, of any account which included a full range of activities. Imagine accounting for the simultaneous activities of washing the dishes whilst having a conversation with a friend whose own simultaneous activities include talking on the phone (perhaps in another language) with another friend whilst also conversing with the person doing the dishes (perhaps a meeting between the three people is being set up, or a visit to a cinema, or whatever). Reference was made to the possibility that within the conception of the non-linear formalism a full account of complex linguistic behaviour could involve many hundreds of tiers/ half-planes. Add in the complexity of managing the non-linguistic behaviour and the scale of the representational challenge becomes clear. Indeed, as we will discuss at the end of this chapter, the significance of the model is the general conception, not the working details of any specific model. The functionality, which is after all the main focus, can be *illustrated* by a model but it's not dependent on it, and the *conception* of how it is that the functionality is delivered is ultimately what matters. We will consider first a specific model, before returning to the general discussion.

The formal scheme of notation from non-linear phonology can be taken to imply a more complex functional model, but pencil and paper and simple 3D diagrams become unworkable (as is obvious from scrutiny of Figure 5.7) – so how can a useful model be made, or must the conception remain simplistically illustrated until the neurological structures are deciphered? This is important because it suggests that complexity of expression of functionality might inevitably invoke the need to consider the wetware – but this is something we wish to avoid. We will consider below a computational model of the functionality which does not oblige us to consider the wetware.

2(6).11.1 The complexity of the model derived from the formalism goes beyond the simple bulk of it. Half-planes traffic data to and from the data-structure referred to as the spine, and likewise with the background processors (BGP) – working "off-line" so to speak. Half-plane processing involves reading other data (in other half-planes, or in adjacent entities in the same half-plane) as well as contributing values as new data. The background processors interact with each other in respect of both supply of data to the half-planes and use of data supplied by the half-planes. One consequence of this interactive complexity is that the whole scheme is bi-directional in the sense that (a) incoming data in

© KONINKLIJKE BRILL NV, LEIDEN, 2017 DOI 10.1163/9789004342996_010 the spine from the senses work their way through to the background processors via a succession of half-plane labellings (so a stream of sound, say, is processed to yield specific speech relevant labellings; and these in turn are identified as syllables, morphemes, words and etc., all the way through to propositions, and with an appropriate multiplicity of half-planes but without hierarchy); and (b) outgoing data are provided from the background processors to half-planes, at a range of scales of analysis all the way through to control of the articulators with fine timings.

2(6).11.1.1 The bi-directionality operates even when the overall task is, say, "speech input", so that an imperfect "recognition" or labelling of a syllable (say) can be "corrected" or influenced by labellings coming through from larger scale structures created in consequence of processing conducted thus far. In more conventional jargon, speech input is not entirely "bottom-up", as there is simultaneous "top-down" contribution to the input processing (but note – this is not analysis by synthesis – see Clark, 2013). In the terminology favoured here, de-sequencing is informed by conjectured larger scale entities supplied as labellings in the data-structure (spine) with a sort of "look-ahead" potential which can be construed as sequencing (and which would be found in sequencing for output). The bi-directionality occurs simultaneously; a creature is constantly engaged in both sequencing and de-sequencing, and in some circumstances both are involved in specific behaviour, such as speech communication, so that at any moment the observed behaviour reflects emphasis or predominance of one over the other. We will explore this aspect of the notional model in more detail in Chapter 7. We can, however, look at a simpler notion of bi-directionality where the sequencing and de-sequencing are uncoupled and taken as separate activities, and we turn now to consider briefly a computational instantiation of this bi-directionality.

2(6).11.2 A doctoral student (Iles, 1995) produced a system which took as input a string of characters – text – and produced as output time varying control parameters for a speech synthesizer – speech. The text-to-speech conversion conception involved a number of novelties in speech synthesis. However, for our purposes here the particular novelty we are interested in was that the overall system structure or architecture was derived from the non-linear formalism, in the manner outlined above. The system was built with background processors for aspects of speech such as morphology and syntax, and the text string was de-sequenced into entities of large scale (morphology, syntax), set alongside the representations of words and letters. The whole complex representation was then re-sequenced as speech output via the synthesizer control parameters (instead of muscle control signals). The synthesis was managed in a non-segmental manner with articulatory target specification (handled by deriving elsewhere and then using in the model a set of pseudo-articulatory control parameters for the synthesizer). This enabled drunk/ slurred speech, for example, to be synthesized at will, through modification of the fidelity with which each articulatory target was achieved. The representation of speech as continuous activity, rather than segments, was important in this project. The overall goal was achieved – a proof of concept for the de-sequencing and sequencing structure of the model, using the spine as a core data-structure. The system was thus completely unlike more conventional "grapheme to phoneme" systems or concatenative synthesizers using stored word renditions or other units.

2(6).11.2.1 The engine for de-sequencing and sequencing was called, in this work, Pantome (for "every way of slicing"). Details can be found in Iles (1995). Constructing it raised a number of questions relevant to operation of the model in general. Two in particular are of note. The fact that Pantome is only a fragment of a (de-) sequencer means that it is not embedded in a larger more general system and thus it needs to be told when to start and stop its work. This is equivalent to triggering the system to work on the data-structure when the input to the spine is declared to be complete. This is handled by terminating the input string with a code-word the presence of which triggers the rest of the engine to process the data in the spine. In reality, of course, there are always other such triggers to indicate that processing should cease or continue, and with what emphasis – as we will discuss in Chapter 7 (you could envisage that stopping an organism's (de-) sequencer is to anaesthetise it or kill it).

2(6).11.2.2 The second aspect of the Pantome instantiation of the model which needs consideration is error-handling. Mislabellings of entities in the spine need to be unpicked without having to unwind the whole data-structure back to the point of error. Or, in other circumstances, the error needs only to be flagged but otherwise passed by (this was not built into Pantome, but is offered here to clarify the main point). The use of clusters of half-planes with *primus inter pares* "representatives" for interworking with other half-planes and clusters means that a correction to data in the spine can be localised within a cluster without necessarily requiring re-writing of the links throughout the data-structure. For work with speech recognition the model would need the capability to flag but ignore an idiosyncratic production, as distinct from correcting an error.

2(6).11.3 Another way of thinking about this point is to consider the general impression given by the formalism and diagrams like Figure 5.7 The reader could be excused for having the view that the half-planes are all linked to each other via the spine – regardless of the scale of the entity/unit under consideration. Superficially this seems attractive – but trying to build a model that actually functions shows us that it is not workable. And indeed it is not necessary – the *primus inter pares* idea, where one or another half-plane "stands for" a cluster of half-planes involved with a unit (say, the syllable) provides something akin to modularity and isolation of perspectives on the data-structure in the spine. The modularity is not comprehensively isolating because the background processors communicate amongst each other, but it suffices to let processing continue with locally contained errors or unclarities in ascription of values to portions of the spine. Additionally, this notion of modularity permits idiosyncracies to be handled as such, with speaker identification provided alongside speech recognition. Similarly, emotional state can be processed alongside the speech processing. Note that my use of the term modularity is not the conventional use which is that language exploits an isolated/ specific module in the brain (see 3(9).18.1).

2(6).11.3.1 Referring back to Figure 5.5, reproduced below as Figure 6.1, we see that a cluster of half planes at one scale of analysis can be "represented" by just one at a larger scale of analysis. This scale independent scheme of representation can continue through to ever coarser scales of analysis. However, one point does need to be kept in mind. It is convenient in this account of the working of Pantome to assume just one representative half-plane deployed in larger scale structures, and so on through ever more coarse structures. The simplicity in the Pantome scheme of things suppresses the need to consider the possibility that at any one scale of analysis there may be more than one *primus inter pares*

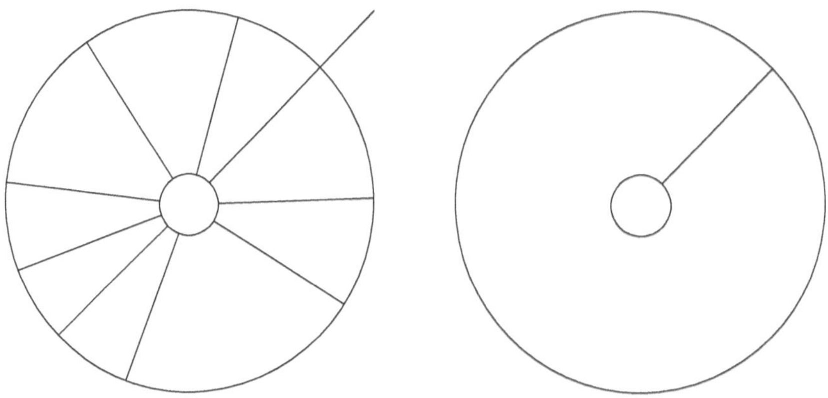

FIGURE 6.1 *Sketch of spine "end-on" showing how one half-plane extends from the spine further than the rest and can become representative of the rest in the sense* primus inter pares.

representative half-plane. So, for example, syllables in speech might be represented elsewhere in structural accounts (say, for stress) by the syllable nuclei (vowels, typically), elevating one phone to the status of *primus inter pares*. This makes sense of the ease with which we find rhymes for words (spoon, moon, June...). Yet, this appears to be over-ridden, or perhaps just accompanied by, another half-plane representative, in the case of other sound patterning possibilities – alliteration, for example "the great grey-green greasy Limpopo river". And even more remarkably, each syllable initial sound can be isolated from its nucleus, "on the fly" by children and adults who can speak unrehearsed *pig latin,* also known as *ig-p-ay atin-l-ay*. The implication of such speech processing skills is that at will people can designate other aspects of speech to be the representative half-plane *primus inter pares* (or even just one of several). The structures are both potentially more complex, and less fixed, than might at first seem necessary. The key is that structuring happens, with certain benefits for general processing, not that it is rigid or immutable.

2(6).11.4 As we move towards consideration of the way real brains serve the Sequential Imperative – and any brain is going to be much more complex than the simple model in Pantome – we need to remind ourselves that the overall scheme is not hierarchical, despite the convenience of using tree diagrams in structural accounts (recall the discussion in 2(5).10.3.3/4/5). This links back to the discussion above about the representative half-planes in the structures; if these are not fixed but can be redetermined "on the fly" then the convenience of thinking about tree-structures becomes instead an encumbrance. Indeed, it could be the case that the best way to think about the tree-structures we think we see in the data we study is that they are our interpretation of the linearization of more complex three dimensional structures. Just as it might seem that they have convenience so it might be that they are artefactual by-products of thinking about the (de-) sequencing processes. They are thus, in some senses, doubly misleading – cognitively convenient but inaccurate; artefacts rather than "the real thing" in the underlying processing.

2(6).11.4.1 There is one particular aspect of structural accounts of language behaviour which deserves special commentary. Much has been made of recursion in linguistic structures – it is thought that recursion is found in such structures but nowhere else in behaviour, and it has a special place in the history of linguistic theory. For example, recursion permits the creation of unlimited numbers of unlimited length sentences in any language. Despite the fact that recursion is not actually found that readily in language behaviour, being instead a property of the theoretical accounts of potential data, it remains an important idea which "challenges" or "interrogates" alternative proposals.

In this context what is required is both a better account of real data, and an account of why recursion is not an issue.

2(6).11.4.1.1 Regarding the accounts of data it has been proposed by Levinson (see for example Levinson, 2013) that in fact recursion – interpreted as embedded structures, not tree structures – can be found most reliably in pragmatic structures in discourse and thus outside the formal theoretical accounts of syntax from which the notion springs. This proposal is controversial.

2(6).11.4.1.2 Recursion is not an issue in the account developed in this book because the rule governed creation of strings of symbols (which provides the formal route to accounts of recursion – see for example Berwick and Chomsky, 2016) is not the mechanism proposed for creating language output. The complex interleaving of multiple and varied specifications of moments of time – creating the sequential data structure in the spine of the system sketched in Figure 5.7 – can create arbitrarily embedded structures such as those described by Levinson (2013), as readily as the morphological structures in Semitic languages, or the endless chaining of prepositional phrases in English ("The cat sat on the mat, by the door, of the house, on the road, near the stream, through the village, in the valley…").

2(6).11.4.2 The reality of using models such as Pantome is that although they are amenable to computational expression – they can be made to work – they are too simple. The issue is not just that representational devices like tree-structures, or multiple half-planes around a spine, or whatever, are convenient. It is that they reduce the complexity of the representational challenge to manageable but inadequate scale. The nervous system of an organism such as a bee contains close to a million neurons, with perhaps a billion connections. Bees can do various things (see Gould and Gould, 1995), including (a) "communicate" about food sources (both distance and direction relative to the hive) in a way which is sensitive to the time of day; (b) fly to the specified food source and back, perhaps as much as 10 kilometers – or roughly a million times further than the bee is long; (c) navigate to the food source and back. However, although the simplicity of a model like Pantome is beguiling and thus perhaps suspect, it is ultimately not unhelpful – unlike the simplicity of computational models based, for example, on "neural network" models of the wetware. One value of the conception underpinning Pantome is that it is obviously scaleable, so even if we cannot produce a model of something as complex as a bee, we could construct a large scale version in a computer, containing perhaps thousands of half-planes. Also important is that the conception can express the way the Sequential Imperative is served; even in a simple model the bi-directional property is obvious, as is the way both sequencing and de-sequencing provide some "look-ahead" processing (this will be discussed further in the next Part).

2(6).11.4.3 One remaining aspect of the possibility of building a computational version of the non-linear phonological model – Pantome – is that it could be used as a computational formalism. With a computer of sufficient power, and software of sufficient elegance (with a user interface to match), Pantome could be used not just to perform as an AI engine in cognitive modelling, etc., but as a 3D notational formalism for capturing and expressing aspects of activity and perception (and attention – see the next Part). This has not been attempted in any significant way (a good interface was available in the version of Pantome built by Iles, 1995).

2(6).11.5 Two topics need coverage for completeness in this Part. The first of these will resurface later, but needs mention here to head off a train of thought that is in danger of distracting the reader. The word "representation" has been used and may provoke in the reader ideas about cognitive representations. Notice that in the earlier discussion of the GCPs (Part 1) the word "entity" was used, along with "cognitive entity". The cautionary note being sounded here is that we need to steer clear of ideas which imply that the representations we can envisage through the operation of the (de-) sequencing engine are in any sense interpretable. The complexity of the "formalism" we are describing as the operation of the computational rendering of the functional model is not only not easy to imagine, it is not interpretable beyond the generality of the "curious sort of time machine" referred to earlier (Introduction to Part 1). The topic is taken up again in the next Part.

2(6).11.6 The second topic which needs mentioning, if only to set it aside and move on, is cybernetics as an alternative conception of brain functionality. This relates to homeostasis but also to computational modelling. The cybernetic notion is that the brain is fundamentally a control system with feedback loops. The idea was popular for a while in the early days of computing, when analogue computing machines were still part of the syllabus in electrical engineering degrees. Since that time the (digital) computational metaphor has become much more popular, with the brain seen as an information processor (as discussed in the *Introduction*). The cybernetic model of the brain and its functionality has little credibility largely because it became too vague (see Frankish and Ramsey, 2012) but also because it fails to explain anything, even metaphorically. However, Clark (2013) attempts to rehabilitate the value of a cybernetic perspective through a focus on the importance of prediction and error correction. Homeostasis might seem to be the best territory for its application, but the problem there is that much of the work to be done is biochemical, not cognitive. Fear, for example, might trigger a flight or fight response, with an adrenaline rush, which pumps up the heart rate, and etc., but the cognitive element is little more than a learned response. Fearless robins have already been discussed (1(3).7.4.1.4). They learn to fear other creatures, such as humans, after a while, but this poses a problem because learning is a difficult story to tell within a cybernetic framework without it all sounding like behaviourism. (See also Chapter 1, FN5.)

2(6).11.7 The overall goal in this mid-Part of the book is to convey to the reader, with detailed illustration, the central idea that one of the formalisms developed for representing structures in speech and language can be generalised to provide representation of other structures in language, and indeed other aspects of behaviour more generally. This last point remains unelaborated so far – details will be found in the next chapter.

2(6).11.7.1 We saw in Part 1 an account of the Sequential Imperative dealing with behaviour in a rather general sense and with equally general accounts of data. The broad-brush approach there served to introduce and demonstrate the SI without fine detail. In this Part we have looked in detail at a specific conception derived from non-linear phonology – real behaviour, with real data – but with the recognition that there is a complication. Attempts to produce representations of this conception as general purpose quickly run into the problem of scale – a real brain, with real processing, is massively more complex than a simple 3D diagram. Even envisaging a computational model becomes difficult as the size and density of the model increase. The reader has to form a general conception based on the phonology, and then scale that up to cover a multiplicity of co-occurring activities. The next Part, Part 3, begins with this larger scale conception but revisits some of the earlier discussion of learning and attention.

2(6).11.7.2 An additional comment on the road travelled thus far is worth keeping in mind. The “serving” of the Sequential Imperative was presented in Part 1 as somehow simply requiring sequencing and de-sequencing to be accomplished by the brain, for the organism. This was then elaborated into two notions of the brain “serving” the SI – both providing for the SI and actually being an engine to make it all happen. In this conception the engine for (de-) sequencing is the active part, with the more passive storage part providing a repository for de-sequenced entities, and the source of to be sequenced entities. And now, following the introduction of background processes (BGP) we see the need for off-line activity away from the (de-) sequencing engine – activity which serves to provide for a more complex notion of the "internal world" which requires the SI, and which makes the notion of serving the SI still more complex. We will address this in the next Part.

2(6).11.8 The possibility of creating a working model of the formal idea is appealing, and instructive. In particular the effort required highlights the simultaneous bi-directionality of the overall system or process. It can be called a "formalism", but that suggests it can be written down. It is a representational scheme for the serving of the Sequential Imperative. We will use the conception in later discussions.

2(6).11.8.1 Despite the simplistic nature of the discussion of Pantome above it should be noted that in principle the formalism could be instantiated as a really huge piece of software, with user interface and attendant peripheral hardware. It might also be worthwhile considering whether there are any limits to the eventual size of a Pantome-like application for modelling behaviour. This would be a major AI undertaking. Such a possibility holds the door open to considering as valid a reductionist account of mind; we will return to that in Chapter 9. AI modelling could influence conceptions of the evolution of the (de-) sequencing engine in the brain itself, and how brain size does (or doesn't) matter (we discuss these issues later).

2(6).11.8.2 However, the point about the nature of the formalism can be considered from another perspective. Providing an account of the totality of language behaviour is not within the bounds of any one written formalism currently. The reality is that this might only be possible in some sort of software system capable of dealing with the dizzying array of half-planes, the BGPs and the rest of it. In that sense the (computational) formalism offers a style of description which is unified across the range of language behaviour, and other human and non-human behaviour, without submitting necessarily to statements in the form of rules (cf. the earlier discussion of the morphophonemics of the English plural, possessive and etc.), or tree-structure diagrams and so forth. The conventional linguist might consider this to be a step backwards. The Cognitive Scientist considers the progress is real – a step forward. However, as the diagrams offered earlier show, it is possible to work with the right sort of diagram of just a small part of the linguistic behaviour being considered.

2(6).11.8.3 Perhaps the simplest summary here is that Cognitive Science now offers a way of doing Linguistics. The reader should note we will now assume Cognitive Science does (or can do) what is required to satisfy Cognitive Linguists, whose motivation is now overtaken by the development of a comprehensive model. At this point, in fact, I would argue that we have achieved a Cognitive Science account of the possibility of language behavior. This account of the possibility of language behaviour is so comprehensive that a specifically Cognitive Linguistics approach is no longer warranted. Linguistic theory still requires development, however, but from a new standpoint. Such theory is not offered here.

2(6).11.8.4 Chapters 4 and 5 (especially) show complex continuous language activity can also be cast in complex segmental detail which nonetheless has to be learnt. The complexity of structuring seems to be uniform for many different aspects and scales of language behaviour, only some of which are expected to be determined by Universal Grammar. It appears unlikely that language has to be treated differently when so much of what goes on is indistinguishable from general behaviour and perception in humans, and indeed in other animals. Whilst it may be that some will see the need for more comprehensive demonstrations of pan-species utility of the model, I would argue that in fact it is now incumbent on the proponents of UG to argue their case more comprehensively by demonstrating (rather than just claiming) that general cognitive principles cannot suffice for language behaviour. We noted earlier (1(3).7.8.1) that it is necessary for those proposing UG to explain how language must depend on the Sequential Imperative (as does all behaviour) and yet at the same time cannot be completely accounted for within the complexities of the Sequential Imperative and the GCPs which are required to make it all work.

In this Part we have seen something of the *detail* of what has to be learned and controlled to make speech work. The account is set in the context of the SI and the GCPs and again the UG proponent needs to consider how it is that the SI and the GCPs can deliver so much detail but still UG is required to furnish something more (and this extra "something" must at the least not disrupt the working of the SI; indeed it has to comply with it). The general implication is important: innate knowledge is not compatible with the SI.

2(6).11.9 We turn in the next Part to consider the GCPs 4–8, as noted in Part 1, and also to consider some of the philosophical implications of the Sequential Imperative. The model discussed in Part 2 also has implications for understanding the evolutionary development of creatures with brains – and these are explored in the next Part.

2(6).11.9.1 GCPS 5 & 6 deal with management of the Sequential Imperative, and this is expressed there through the use of the term *attention*. This feels right in the sense that one is aware of attending to sensations and actions, both for themselves and in respect of decisions made and alternatives considered. The subjectivity is problematic if pushed too far, but helpful for orientation towards new material. There is an alternative perspective on the material of the next Part covering these GCPS; the scheme set up so far is uncontrolled or unmanaged. The reader really needs to focus on the notion of system management (in engineering terms, perhaps, and thus redolent of the Pantome model) as eventually giving substance to the notion of attention. However, the treatment of management of the SI in Chapter 7 inevitably points ahead to ideas to be covered in Chapter 8 so there is a problem for the reader (and the author) in controlling both vagueness and repetition. In particular, in Chapter 7, when dealing with the management of the SI we will list the problems and then move the terms of the discussion over to illustrations from an attentional and subjective perspective.

2(6).11.9.2 However, before starting the final Part it is worthwhile reviewing the totality of the story so far and seeing what big themes carry over into the next Part. The account of the Sequential Imperative, and how we get from making tea to learning how to make unexceptional plural forms in English, is a story of increasing complexity in behaviour, with increasing sophistication of the account on offer. The reader might like to refer back to the simple diagram Figure 2.2 and consider how that could be used to sketch an account of tea making – and then do the same using the diagram Figure 5.7. If that seems like too much hard work then perhaps some reflection on the notion of segmentation in behaviour will be more effective in convincing the reader of the utility of the general insights. The insights from phonology really do generalise to all behaviour. The two Parts A and B present the top-down view from general Cognitive Science (A) and the bottom-up view from Linguistics (B). Here, at the end of Part 2, we see that these perspectives can be combined in terms of a model derived from phonology. The overall conception is used in the final Part, with the model illustrating yet more complexity in behaviour. It is already clear that trying to draw the diagrams showing multiple half-planes linking to a spine is beyond pencil and paper. It may be workable in a software system interface to a computational model, as was the case with Pantome (but even that was very simple).

2(6).11.9.3 A theme which recurs is that of the syllable – a major factor in the discussions in this Part it resurfaces in the final Part. There is something appealingly obvious about syllables and the discussion in this Part has been easier because the nature of the syllable has been a convenient way to introduce ideas. Some of that simplicity carries over into discussion of the evolution of behaviour we call language.

2(6).11.9.4 Our last point in this Part is simply that it now seems safe to assume that the complexity we see in the *expression* of functionality, and which seems to be readily captured in a computational model such as Pantome, is not in fact complexity in functionality *per se* (see 2(6).11). In turn this means that we do not need to consider the wetware when we are trying to understand what the brain does – what it is for. If the reader is unconvinced I can only suggest the doubts are set to one side until the rest of the material is covered!

PART 3

Behaviour and Evolution – on and off Planet

Introduction to Part 3

In this Part – the third and final major Part of this book – we need to build on the material of the previous two Parts in ways which carry the ideas forward to cover the implications of the Sequential Imperative for understanding the structure of behaviour. The general concerns of Part 1, and the Cognitive Principles discussed there, give us the "top-down" Cognitive Science view on behaviour. Part 2, in contrast, deals with lots of details of the organization of one particular aspect of human behaviour – speech – and provides a "bottom-up" view of the realities of behavioural organization. In Part 3 we find ways to combine and generalize these two perspectives to illustrate the complexities of the organization of behaviour whilst also elaborating those complexities further. We extend the account to cover evolution, inner speech, consciousness and the possibility that we might already understand something of what it is like to be an Extra-Terrestrial Intelligence. But all the while we are anchored by the Sequential Imperative.

Chapter 7 offers more discussion and detail of the GCPs 4, 5, & 6. Learning and attention, along with the use of the model set out in the previous chapter, are discussed. The main topic is management of the Sequential Imperative, and the detailed coverage extends the work of the first two Parts to consider, in detail, what has to be done to deliver on the statement that the brain serves the SI. Chapter 7 also provides an introduction to thinking about language from the perspective of the SI.

Chapter 8 covers some work on evolution and language – not in a compendious way but in relation to the pan-species significance of the Sequential Imperative, even though language is uniquely human. The key idea developed here is that language (or indeed any other aspect of behaviour) does not evolve. Neither does the functional specification evolve. What does evolve is the (de-) sequencing engine which can be described in terms drawn from Chapters 5 and 6. The interest is not the wetware, it is the delivery of functionality. This perspective contrasts both with the conventional view from Universal Grammar, and the specifics of some work done on the importance, for "evolution of language", of the rôle of the syllable and mandibular cyclicity in movement. The heterogeneity, or otherwise, of populations of language users also turns out to matter.

Chapter 9 takes up the issue of Nagel's question: "What is it like to be a bat?" (Nagel, 1974). The ubiquity of the Sequential Imperative gives us a basis for developing some thinking on this topic in the context of GCPs 7 and 8. We can consider some aspects of cognition to be universal and this allows us to reflect on consciousness and thought in other species including, perhaps, Aliens.

© KONINKLIJKE BRILL NV, LEIDEN, 2017 | DOI 10.1163/9789004342996_011

The focus on management of the Sequential Imperative, the topic of Chapter 7, our next chapter, is important because the dizzying complexity of both the data and the model described in the previous Part is such that sequencing and de-sequencing cannot be left to run without some sort of control; the result would be behavioural and perceptual chaos. The need to make the SI deliverable reveals complex problems of management (but no manager homunculi). These are discussed first in "systems" terms, and subsequently in more subjective terms.

In Chapter 7 we will also return to the perceptual and behavioural asymmetry not addressed in GCP3 (see Chapter 2, FN3). It seems as though GCP3 is "designed for" hearing and speech which demonstrate complementarity not obviously provided in the visual world. Indeed, the original impetus for formulating GCP3 came from study of non-linear phonology, but the claim is that GCP3 is not modality specific, and the fact of signed languages shows us that the dizzying complexity we find in speech is not modality specific. We will look at this issue again later (3(7).13.4.4.3). We also consider in some detail the issue of learning – learning language specifically, but not exclusively.

We need in this next chapter to reflect on another general point. The model and scheme of explanation we are using is based on speech production. Pantome is an instance, but the approach is more generally based on non-linear phonology. This does not mean that speech capability is ubiquitous. There are in consequence some pitfalls which need to be explicitly mentioned so that the reader can avoid them. Firstly, we need to note that although the model is based on speech this does not imply that all brains are capable of processing speech or human language. There is a drive to try to show linguistic capability in other species (see de Wall, 2016:95 ff.) but this is not supported by the approach taken here. Secondly, there is the whole question of the rôle of language in cognition. Cognition in other animals is, we presume, language free. Is language necessarily involved in human cognition? Again, the model does not force that issue and the topic is taken up later in this Part.

CHAPTER 7

Management of the Sequential Imperative

3(7).12 We begin with the need to tie up some loose ends concerning GCP4. Learning, especially when we consider a child's acquisition of language, is a difficult topic to extract from pre-conceptions and theoretical conundra. But in fact learning is a bigger problem than just the achievement of language by children, and some consideration will be given to these wider issues. Within the bigger picture of behaviour controlled by the Sequential Imperative learning is seen as serving the SI by providing an increase in the variety of ways in which the SI can be delivered. Learning makes possible more behaviour. We then turn to a major topic – attention. Attention is defined by the model we have developed – it is about management of the way the brain serves the SI. Because the management of the SI is complex we approach the material in a sequence of subtopics. But first GCP4 and learning, especially learning of language.

3(7).12.1 What is it a child learns when they learn a language (mother tongue, not a second language in a school setting)? A general, but not universal, view has it that the crucial component learned by the child is the set of rules used to produce syntax. This can be widely interpreted (in the case of a child learning English) to include learning the correct form of past tense for verbs – children apparently go through a phase of "over-generalizing" a rule's application, to give "go-ed" in place of "went", for example (see Berko, 1958). There is, however, a problem with learning rule-governed behaviour in syntax. Any such behaviour under-determines the rule-set used to produce it. If a child encounters the product of adults' use of rules to generate language then those rules (as rules) cannot be completely recovered from observable language behaviour. A simple illustration is provided by the logical rule "If P then Q"; less abstractly this can be given the form: "If Ham then Eggs". The logical possibilities are that in addition to the core case, one could have no Ham, in which case whether or not you get Eggs is unspecified – you could, but then again you might not. What is not permitted *by the rule* is having Ham without Eggs. Because it is not permitted it never happens, and the child learning about Ham and Eggs is given no information about this impossible outcome. There are of course ways around this. The child could ask – "can I have Ham and no Eggs?" and be told that such an outcome is not valid. Or maybe the information is volunteered by the provider of the Ham and Eggs. Essentially, the missing outcome in the data

© KONINKLIJKE BRILL NV, LEIDEN, 2017 | DOI 10.1163/9789004342996_012

experienced by the child has to be provided as metadata – information about the case – because the case itself ("Ham without Eggs is not permitted") is never generated by the rule which simply fails to generate Ham without Eggs. And without the extra information the full specification of the rule, *as a rule*, is not possible. A child could say that she has never experienced Ham without Eggs, but that is not the same as saying that according to a rule she cannot ever experience Ham without Eggs. If she is not supplied with the labelled negative data, the metadata, the child cannot formulate the rule. And so it is with syntax – children cannot formulate the rules used to produce valid utterances because the labelled negative data are not encountered in the child's linguistic environment. Grammar, as a system of rules, is under-specified or under-determined by its output.

3(7).12.1.1 There are three ways around this problem – the language learnability problem. The first is to say that the chain of reasoning provides an *argumentum ad absurdum* demonstration that children do not learn language as a set of rules (and thus, incidentally, adults do not use a set of rules to produce language). Their behaviour can perhaps be described as if they were following a set of rules, but that is quite another matter (cf. the argument in 2(5).10.2.3.2 about phonological "rules" as descriptions). The second way around the problem is to say that they do indeed use a set of rules, but the labelled negative data – the metadata – are more readily available than currently understood, or may be induced somehow, or they are pragmatically sign-posted, or whatever, so that children can indeed formulate rules. The third way is to develop a notion of innate language capabilities which make up for the lack of negative evidence. The information deficit (lack of negative evidence) is also known as the "poverty of stimulus".

3(7).12.1.2 The innateness solution has been quite popular with some linguists for a few decades, but is falling out of fashion. There is in any case an inherent problem with it. The innate capabilities have to be brought to bear on data at the right time, to "compensate" for the deficit. These capabilities are generally referred to as the Universal Grammar (UG) and within each child, as language is learned, various settings or options in UG have to be triggered appropriately so that the UG is then configured for the language being learned. Berwick and Chomsky (2016:68–69) put it this way:

> The [Principles and Parameters] approach is based on the assumption that languages consist of fixed and invariant principles connected to a kind of switchbox of parameters, questions that the child has to answer

> on the basis of presented data in order to fix a language from a limited variety of languages available in principle…

The difficulty is that the linguistic properties of any trigger, for it to be able to function as a trigger, have to be processed by the still untriggered UG. But at the same time this processed material must be conformant with the post-triggered UG for the properties of the trigger to be recognizably relevant and thus for the triggering to happen (the trigger is only specifiable in post-triggered terms, otherwise the whole triggering enterprise is pointless). An example might be that a language being learned places adjectives before nouns. The UG will need to be triggered for this (as opposed, say, to placing adjectives after nouns). The notion of "adjective before noun" as a structural property of the language (rather than an isolated case, which could be an error of processing) has to be recovered from the data processed by the untriggered UG – but this is not possible as the grammar at that point cannot formally differentiate the two cases.[1] Alternatively, the untriggered grammar can be supposed successfully to label the data as an instance (of the structure) of "adjective before noun" in which case the UG is not required. So if we try to promote the value of UG we must recognize that the child's processing of the incoming linguistic data has to be capable of producing the fully specified trigger data, but then having done so the UG becomes irrelevant to the learning process at precisely the moment it might be used (see also 2(6).11.8.4).

3(7).12.1.3 The innateness hypothesis has been criticised on other grounds (cf. Putnam, 1971). My own concern is that it is an example of a mechanism or model derived in part from a false premiss which could be labelled "The Wonderment Premiss", or perhaps "The Incredulity Premiss". The point is that just because the child's achievement of language seems so amazing, remarkable, fast, incredible, awesome,… and because the structure of language itself is so wonderful, complex, rich, powerful,… she needs some help along the way to make learning it possible. The invocation of "Wonderment" or "Incredulity" in an argument is surely better understood as flagging the need for re-description of the problem in less flamboyant terms (authorial licence notwithstanding) and certainly less complex terms formalistically. So maybe it really is simpler to assume children learn rich and complex behaviour without the need for innate assistance with rules. And indeed, reflection on the complex phonological and

1 Berwick and Chomsky (2016:68–69) discuss the ordering example noting that the actual order of words in the output is not the issue, rather it is the underlying structural specification. The force of the complaint here is not removed by this more nuanced account.

morphological patterns discussed earlier, along with the complex muscle contraction patterns required to produce speech, suggests that if the child can do all that without innate help (the equivalent of UG is not proposed for these aspects of speech/ language), then why is syntax special? The answer, of course, is that it is not (cf. 1(3).7.1, and see also 3(8).15.4.1).[2] Likewise, and in passing, it should be clear that accounts of behaviour based on the following of rules which have to be learned (somehow) don't appear to be required either. Descriptions of behaviour which are more or less readily formulated as rules (and not all behaviour is so easily described) should not be used to imply that the behaviour is indeed the result of running rules on some sort of engine/ computer.

The reader should recall the discussion of VOT in 2(4).8.1.2.5. The child learning English needs to hear, and process as invariant across native English speakers, a distinction of ~40 milliseconds in voice onset time (~15 ms for unaspirated stops; ~55 ms for aspirated stops), in regular complementary patterns of occurrence (where one is heard the other isn't and vice versa). And then she has to learn to produce that contrasting pattern of occurrence. And all of this is done without any awareness that it is being done and without any parental or societal "corrections", and possibly in the presence of non-native speakers of English for whom the distinction is not relevant or linguistically "valid". This is managed by children *en passant* as they get on with learning English.

3(7).12.1.4 GCP4 states that learning serves the Sequential Imperative. The claim is that because the SI is unavoidable whatever a creature learns will unavoidably be involved in delivering or serving the SI. What else could learning achieve for an organism? (Yes – this does seem rather bleak but in fact is not, as we will discuss later.) When looked at this way the rôle of learning alongside the inevitable innate abilities in some creatures produces complex behaviour expressive of the totality of the Sequential Imperative (both the "off-line" aspect and the (de-) sequencer). Why is innatism acceptable here but not in respect of language learning? Two reasons: (a) innatism is not being invoked to replace apparently inadequate learning processes or stimuli; (b) innatism is assumed to be basic and more like scaffolding for complex learned structures. If a foal can get up and walk shortly after birth this does not rule out subsequent learning of gait or gait modifications; the former reflects innate ability, the latter learned skill. Likewise bees can navigate (innate) but where they go, and how far, are not programmed, even if the components of the "bee language" – the "waggle dance" – are innately supplied. Similarly,

2 For an infant, learning a signed language is as seamless, wonderful, awesome, etc., as learning to speak: modality is not an issue.

some birds (e.g. the Wheatear (*Oenanthe oenanthe*) and Greenland Wheatear (*Oenanthe oenanthe leucorhoa*)) migrate huge distances (up to 9,000 km twice per year), and sometimes thousands of kilometres over oceans, presumably on the basis of innate navigation abilities. The migration route might conceivably be thought of as innately programmed, but it perhaps makes more sense to assume that it is learned from other birds making the same journey.

3(7).12.1.5 One aspect of learning which needs clarification concerns the atemporal nature of cognitive entities. Part of the core idea of the Sequential Imperative is that the brain contains atemporal cognitive entities which are cast and recast into sequencible forms "all the way down" to execution as behaviour (and conversely, the sequentiality is stripped out "all the way up" from sensory input to cognitive entity). Cognitive entities in the background – "off-line" – are changed by learning, and thus active in a sense we have not really acknowledged earlier (but see 2(5).10.3.8). This aspect of learning doesn't undermine the overall notion of the Sequential Imperative because the temporal characteristics of the changes in cognitive entities through learning are not aligned with temporal structures in the "external world". Likewise, the fact that entities may be changing in the background, as the result of internal cognitive processes, does not mean that the SI is undermined. We will revisit the issue of changes in the "off-line" entities and their processes, in relation to the capabilities of the (de-) sequencer, a little later. The SI is required regardless of the state of the off-line background – which could be envisaged as in flux in much the same way as described by Lashley (1951). Now we turn to consider in more detail the management of the SI, defined here as attention.

3(7).12.2 The material above strengthens the argument that explaining language behaviour doesn't require a special story to be told. This is not the same, of course, as telling other stories about other behaviour and showing these really can cover language. The overall model suffices to provide the generality, but there are sufficient pointers to other behaviour in other species to allay any concern that I have merely replaced one "special case" story about language by another.

3(7).12.2.1 The material in the next sub-Part covers GCPs 5 & 6 from the point of view of how one could make the model described earlier (and see Figures 5.7 and 6.1) actually work. The brief discussion of the Pantome system (Chapter 6) is there to show that the model can be made to "tick", so to speak. What is now required is a belief that the full-scale model can be made operational, and that this can be imagined somehow. Yet, the system of half-planes at all imaginable scales interacting with the central spine – with entities of all imaginable durations – is really complex beyond anyone's capacity to envisage as a system "running" other than in the most general sense. However, belief in the functional model can be engendered by looking at the management of the system of (de-) sequencing. What is reasonably obvious already is that specific factors need to be controlled – coupling of domains to the BGP, both automatically and through processes of selection; simultaneity (and bi-directionality); differences of scale of management; flexibility/ change in the entities being deployed; and so forth. The discussion will avoid where possible drawing in material from Chapter 8. But inevitably some forward reference will be made, and in Chapter 8 some recapitulation will be required. The material is really new because it is essentially based on the use of the model which itself is new. Working through this material – at the core of the book – may require the reader to take some things on trust, to be returned to later when the bigger picture is secure in mind.

3(7).13 In Chapter 3 – 1(3).7.2 & 3 – and at the conclusion of the last chapter – we noted that GCPs 5 & 6 are to be covered in this chapter. For convenience here again are those two principles:

> GCP5 Attention is the management of the processes of sequencing and de-sequencing.
>
> GCP6 Affect is an attentional mechanism.

The model discussed earlier provides a basis for explaining how attention works – both as behaviour and as part of delivering the Sequential Imperative for the brain-bearing organism. We will consider this now, but in general terms, with minimal reference to complex diagrams such as Figure 5.7. The reader should keep in mind the architecture of the model diagrammed, but needs also to note that although the conception of attention we are exploring is partly defined by the model it does not require detailed structural representation. There is an inevitable risk that generalizing a model in this way (top-down) is not the same thing as demonstrating, with details, its generic value (bottom-up). The reader may have to work through the generalization to the details if they are to feel convinced by the account on offer. Furthermore, the details are themselves somewhat conjectural – yet it is to be hoped that ultimately the sketch-work will seem obviously worth working with in relation to the rest of this account. One other point to be kept in mind here is that a consequence of the focus on speech in Part 2 is that the account below opens up with a human emphasis. However, as has been pointed out repeatedly in this book the general account being developed is pan-species, so we will need to return to that aspect of generalization.

The Sequential Imperative needs to be managed in several different ways, as we will see shortly. We tend in normal usage to use the word *attention* to refer to the processes of selecting or focussing on an activity, sensation, domain etc. My aim here is not to unpick notions of attention and show how the management of SI fits with those notions. Nor am I concerned to redefine existing conceptions of attention based on analysis of the management of the SI and understanding of (de-) sequencing. Rather, we need to sort out the complexities of how it is that the SI must be managed. It then transpires that *attention* seems an appropriate way to refer to much of this management. The GCPs 5 & 6 capture that conclusion in readily understood concepts; to reach the point of obviousness we start out from the need for, and complexities of, management of the SI.

3(7).13.1 The complex model sketched in Figure 5.7 is unconstrained in respect of the flow of entities into sequence, and the flow of de-sequencing. Without management of the SI the whole affair would become dizzyingly chaotic, as noted in the introduction to this Part. Therefore, let us assume that proposing the model as generically important and insightful carries with it the requirement for explaining how it could actually function. The reader could imagine that the functionality of a timeswitch can be explained by analysis of the functionality of its subcomponents – the timer, the display, the setting mechanism, the switch – all of which can be elaborated without mentioning the functionality of the mechanism itself. However, such an analysis of subcomponents only makes sense when integrated by the over-arching functionality. We therefore need to consider how the SI is managed, in detail, in order to appreciate the totality of the functionality of the brain. Inelegantly, this process of reflecting on the details of the management of the SI has somewhat the characteristics of a list.

Notice, incidentally, that if understanding of the processes of management of the SI is achieved then the door is opened to understanding the behavioural, and possibly neurological, implications of various sorts of failure of SI management. We will not pursue this beyond observing that for some this might be a powerful incentive to make the model "deliver", and thus to make understanding of the management of the SI an important goal. Let us now look at the management issues.

3(7).13.1.1 Speech production is an appropriate illustration for our purposes because the flow of entities as utterable sounds is at the same time a flow of entities with syllable structure; a flow of morphemes; a flow of entities described as words; a flow of sentences, and so forth. Likewise hearing sentences, etc. There is much going on in parallel and the structure of language is about co-ordinating or managing that *simultaneity*. In fact, however, language is only special because using language requires management at a variety of scales where there is much arbitrariness and semiotic exploitation activity (including use of context which has to be managed as well). Spoken language, and the material of Part 2, provide rich detail of what the management of the Sequential Imperative must deliver. Other types of behaviour – preparing a meal, for example – also demonstrate complex parallel structures in activity with co-ordinated management on various *timescales*. And, looking ahead a little, it may transpire that *Homo sapiens sapiens* is special only in regard to the complexity of the management of the SI that it can deliver.

This illustrates our first two aspects of the complexity of management in the SI, seemingly related but not necessarily so. Activities need management of *parallel threads* of structuring. Additionally, and only apparently related to the first, activities need management on different *timescales*. The link between these two aspects is that parallel-ness is obviously achieved when the parallel sets of entities differ in the timescales involved (sentences versus words versus syllables versus phonemes), but variation in timescale is not necessary to achieve the parallel threads of structuring (syllables and morphemes, for example, may not map onto each other, but may well be similarly sized in duration).

Speech also illustrates the third complexity in the management of the SI – related to the two just outlined. The *alignment of entity boundaries* is not necessarily precise: a syllable boundary will be the conjunction of boundaries in entities specifying voicing, nasality, frication, etc., but these components will not be perfectly aligned in time (see 2(5).10.2.4.2). Likewise, but more generally, the parallel threads of structuring, noted above, with entities and activities having different timescales, inevitably require some entity boundaries to align in some sense, but not with infinite precision of simultaneity. Anticipatory articulation of one sort or another is frequently found – beginning (in part) one activity (entity) before finishing something else – and this has to be managed. I do not walk to the car with empty hands, survey the door and lock, compute the need for a key, and then make the moves required to extract it from my pocket. No – as I walk to the car I locate the key in my pocket so that on arrival I have it in hand ready to open the door (those with more fancy car keys demonstrate their anticipation by unlocking the car using the wireless remote control on the fob, whilst they are still walking towards the car). This is structurally the same as commencing lip-rounding before the vowel which requires it.[3]

---

3 The Maximal Onset Principle (2(4).8.2.2) may just reflect a processing focus on the "beginnings" of entities/ activities. The car key illustration; the fact that semantic prefixes

3(7).13.1.2 The processes of sequencing also present a management challenge through exploitation of variability provided by *componentiality*. The diagram in Figure 6.1 illustrates the *primus inter pares* notion whereby one half-plane comes to represent others in a domain – indeed as definitional of a domain. What we have seen, however, is that this is not immutable. Different half-planes can be pressed into service as desired. Syllables in speech can be managed in a variety of ways to achieve certain aims, and the speaker can choose as she goes along. Alliteration and rhyme might require focus on various components in the syllables. Children's *pig latin* places similar demands on the speaker who can prepare for external sequencing a string of phonemes and then re-arrange the syllabic constituents, with adjustment to the phonemes (the listener has to unpack the bizarrely delivered speech – we will deal with de-sequencing shortly). My point is not that this is remarkable (or not), simply that there is no fixed predictable "flow" to the sequenced surface. Likewise the choice of vocabulary to describe an incandescent sunset, or whatever. This can all be considered management of the power and flexibility of componentiality. Notice, in passing, that presumptions of hierarchy in structures – say, spoken language structures – can finesse the need to consider some of the management issues.

The componentiality of concern above covers indirectly another aspect of scale. When thinking about management of the Sequential Imperative there is a need to reflect on the size of the brain involved. Bigger brains mean more componentiality (even if only because bigger brains go with more body mass, and this means more behavioural detail to be managed). We are here talking about the scale of the sequencing engine and of domains where half-planes (in our model) are grouped with a representative half-plane as *primus inter pares*. We return to this later (3(7).13.4.2, 3(7).14.1 and in the next chapter).

3(7).13.2 In addition to *simultaneity, timescale, boundary alignment, componentiality* (flexibility and scale), there are other aspects of managing the Sequential Imperative that need to be considered.

3(7).13.2.1 Sequencing and de-sequencing need to be managed independently – that is a facet of SI management in its own right. However, they also need to be managed in coordinated fashion in respect of incompletely specified entities, where the issue looks more like managing futurity. At any moment in time a portion of each of the many entities "at the surface", so to speak, is not fully

are preferred (with syntactic/ structural suffixes); the "universal" which notes that sentence structures are predominantly not Object first; all point to anticipation and "fronting" as helpful for processing – both sequencing and de-sequencing.

specified (if they were fully specified they would already be in the past; we are considering the temporal domain of sequencing of articulation and de-sequencing of proximal stimuli). This incomplete specification occurs regardless of the scale of the entity – but is especially true of the larger scale entities spread out into the future where both sequencing and de-sequencing are involved in providing partial specifications of yet to be externalized (or internalized) entities (see 3(7).13.4.3 below). Sequencing and de-sequencing are in this scenario simultaneously inter-related, providing for top-down processing and bottom-up processing to be coupled bi-directionally in the completion of specifications. *Bi-directionality* is thus an important aspect of the SI, and this too needs to be managed. Remember, however, that *sequencing and de-sequencing* also have to be managed when they are not coupled through the need to resolve incomplete specification (i.e. where the sense of bi-directionality is not invoked: talking whilst watching some eggs cook). This brings the tally so far to 6 aspects of the management of the SI which need to be understood.

3(7).13.3 In this chapter we are in general terms seeking to align the topic of management of the Sequential Imperative with the more conventional notion of attention, although the aspects so far identified do not have very obvious analogues in attentional terms. We can now consider two aspects of SI management which more readily match conventional notions of attention – indeed for the first we can use that word; this aspect is what we can call *automatic attention*. The second is *multi-tasking*. The two are potentially inter-related, as we will see.

3(7).13.3.1 Attention is automatic when the brain-bearing organism manages some aspect of the Sequential Imperative in a way which appears routine or self-actuating in some sense; failure changes the fact of the attention into a task which needs management explicitly. An obvious example in human behaviour concerns maintenance of balance. In human vision automatic attention processing (learned rather than innately specified) provides a supplementary way to maintain balance. Automatic attention can be shown to be operating by arranging for it to be disrupted, and this is relatively easy and indeed may be something quite widely experienced. If you are standing on a railway station platform, talking to a friend, say, alongside a stationary train, perhaps at the end of a carriage where there is a door – and the train suddenly moves with a jerk – the likelihood is that you will sway or totter a bit to secure your footing and thereby your balance. What has happened is that your "visual system" has noted the vertical lines of the edge of the door, the end of the carriage, whatever, and is attending automatically to this indication of the vertical. When the indication moves your system of balance (which depends in part on visual evidence) computes that you have swayed – so you have to correct your movement. And in that process you discover that you didn't move – the "world" did, or at least the visual reference point to which you didn't know you had been automatically attending.

Similarly, and again the account concerns behaviour in humans, the learned behaviour of walking depends on automatic attention (gaits differ; the Brazelton tests on neonates for placing and stepping do not indicate that the complex behaviour of walking is not learned in humans). This is readily demonstrated through the well-known behavioural adjustment to gait known as "getting one's sea legs" which people have to go through when they travel on large ocean-going vessels. You adjust to the motion of the deck which is all the time not quite where you (your feet) expect it to be. Walking is at first rather tricky, but after a while the automaticity returns, but is different so whenever you go ashore you have to re-adjust to the stability of surface encountered there. This automaticity is less inscrutable than visual maintenance of balance, but is nonetheless automatic and quickly consigned to the realm of behaviour one does not have to think about.

Experienced drivers are often worryingly (to themselves, and others) engaged in an automatized set of behaviours (for example, changing gear according to sense of speed and/or engine noise, but without really being aware of doing this). From time to time the demands of the task are such that the driver will know they have to become involved in task management, and will tell a passenger to stop talking, say, so that they can concentrate. At this point attention itself is being managed, not just the task.

3(7).13.3.2 Multi-tasking is the management of attention in respect of a few on-going actions and "inputs" – both sequencing and de-sequencing can be involved. The car-driver multi-tasks but has to manage the attentional demands involved if the traffic and/or road conditions become unusual or unfamiliar. However, multi-tasking is variable across species, according to brain size (see below) and some multi-tasking can indeed be automatic. In fact the human visual balance story recounted above reflects some automatic multi-tasking because the "visual system" was not wholly pre-occupied with keeping the person upright.

Management of the Sequential Imperative is more or less complex depending on the size of the brain concerned. This is not, of course, to say that an elephant has a more complex SI management problem than, say, a cat. It is not absolute brain size, but size relative to the body (larger bodies have more muscles, more complex sensors, and etc., see below (3(7).14.1), and see Chapter 1, FN4). However, suffice to say here that in the service of the SI a brain can exploit some automaticity to achieve multi-tasking, without encountering the need to manage multi-tasking *per se*. By automaticity is meant the use of entities for behaving (perceiving) without the additional "overhead" of managing or selecting those entities alongside the other tasks also being managed. (Organisms with very small brains may not be able to do much that we could recognize as multi-tasking – automatic or otherwise.) For example, a bird can look for food and simultaneously scan the skies for predators. A horse can walk and defecate simultaneously. Multi-tasking matters and has to be effective – what matters more, as we see below, is the management of multi-tasking.

Setting aside issues of automaticity and brain size, it remains the case that multi-tasking may have to be managed explicitly in some circumstances. This feels, in human terms, very like attention – we believe we can attend to several tasks, activities, perceptual inputs, whatever, "at the same time". In the paragraphs above the contrast has been set out between *automatic attention* and *multi-tasking*, but we have noted the possibility that in some circumstances multi-tasking may (in part) be automatic. The clarity will fade further when we consider multi-tasking in more detail below; for now we can consider that these two aspects take our tally to 8. Management of the SI, making it actually work, is complex, but we should not be surprised. It may, incidentally, involve distributed cognition as well as language (this is discussed at length below).

3(7).13.3.3 It was noted earlier (1(3).7.2) that attention can also itself be behaviour which needs to be managed. We will examine this point in more detail shortly. First we need to note that some aspects of management, or organization, of the Sequential Imperative are not themselves in need of management, as we saw above in the discussion of automaticity. In one sense, automaticity is an ability to manage aspects of SI without managing the management, but it is not the only way to think about simplicity in SI management. In terms of the complex model introduced earlier it is as if the "pathway" from Background Processes to muscle contractions (and the return, from sensors to BGP) is able to run without intervention; there is a directness akin to a behaviourist model. All brains will have this capability, possibly including modest multi-tasking and bi-directional processing, as a sort of base scale of SI organization. In addition, necessary multi-tasking for, say, eating and predator watchfulness, is not a matter of choice; it happens. Notice that throughout the discussion of multi-tasking we have not differentiated between simultaneous and sequential (interleaved) multi-tasking. Chewing gum whilst walking demonstrates simultaneously organized multi-tasking; a bird might repeatedly bring food to the nest for chicks, but just before each trip it will pause and look around for predators – the watchfulness and the feeding are interleaved tasks managed in a demonstration of non-simultaneous multi-tasking. In short – attention may switch from moment to moment, possibly within a modest multi-tasking scenario but nonetheless not requiring the management of the SI to become a fully-fledged activity in its own right (with all the expectations arising from GCPS 1–7 to be applicable).

3(7).13.3.3.1 When the material being managed is complex and large scale – multi-tasking with many domains and scales involved *en route* to sequencing for externalization (or indeed from de-sequencing for internalization) – then the *management of the Sequential Imperative can become behaviour which itself requires management*. This appears at first sight to be recursive (an interesting thought) but does in fact have bounds – domains are not indefinitely extendible. Perhaps this is most obvious in the case of managing contextualization (cf. distributed cognition), which is part of attention, alongside managing the behaviour (and perception) which is contextualized. What we are considering, it now appears, is *multi-tasking and its management*.

The complexities of managing the Sequential Imperative need unpicking. Multi-tasking for speech, along with, say, making a meal (and even at the same time as making a meal), requires multi-tasking to be managed; this then constitutes the management of management of the SI (and is the most obvious domain in which to find contextualization). Multi-tasking can itself be behaviour, with entities and componentialities; with domains and *primus inter pares*; and all the rest. Learning is involved, as for other behaviour, and for the same reasons. Referring back to 3(7).13.1.1 the specialness of humans is thus the capacity for management of SI management which is required for multi-tasking in the complex sense we find it in speech and other activities, and which we do not find in (many/any) other species. So, we have managed to slide a distinction between multi-tasking and *managed multi-tasking*, with the latter being yet another aspect of management of the Sequential Imperative. Managed multi-tasking brings the tally to 9. But there is more to consider.

3(7).13.3.4 If attention – management of the Sequential Imperative – can be a behaviour in its own right it comes with componentialities and the deployment of sequencing and de-sequencing activity in accord with the general model discussed earlier. The key concept here is that management of the SI is itself governed by the SI – so attention (because that is what we are talking about) involves entities with sub-entities, and so forth, subject to the same sequencibility constraints as behavioural entities, but with the proviso that externalization is not necessary (but monitoring for structural integrity is done via incoming data and de-sequencing in the normal way). However, externalization may indeed be observable – *management of the Sequential Imperative may be observable*, (tally now 10). Again, contextualization is operating here – context must be recoverable, and in the case of distributed cognition recoverable by others.

Management of the SI – attention – is in fact often observable, especially so when the management is itself complex behaviour. Perhaps the most obvious illustration of the observability is eye-gaze direction. This can be read directly from observation of the eyes of another creature – one has the strong sense that one knows "what they are looking at" or the direction in which they are looking, and one can direct one's own gaze to the same target (one believes). There is no reason to suppose other animals cannot do this to some degree. Relevant, but actually different, is that eye-gaze direction can be triggered by alarm calls. In birds an alarm call indicating an aerial predator could be enough to indicate to another conspecific that whatever else it was doing it perhaps should attend visually to the sky (we will return to alarm calls and attention later, in Chapter 8).

3(7).13.4 In summary, then, management of the Sequential Imperative is complex, with some 10 factors identified as relevant to the successful achievement of the realization of the SI in behaviour (there may be more). That serving the SI is the function of the brain need not be difficult to understand. However, the functional specification which "delivers" on this requirement must cover the following: *simultaneity, timescale, boundary alignment, componentiality* (flexibility and scale), *bi-directionality, sequencing and de-sequencing, automatic attention, multi-tasking, management of the Sequential Imperative as behaviour, observability of Sequential Imperative management.* These are the (sketchy) details of how to make the SI work. Notice, in passing, that some of these might be more evident in human behaviour than in the behaviour of other animals. For example, consider the combination of multi-tasking sophistication and bi-directional coupling of sequencing and de-sequencing demonstrated in the performance of a piece of music from a score (especially if the instrument is a wind instrument and the performer can manage the "circular breathing" technique[4]). This complexity is unlikely to be managed by another animal, even though the general scheme of behaviour is universal. Thus, the list of 10 factors is comprehensive

4 Demonstrations of circular breathing are easy to find on You-Tube: Bora Dugić playing "Zajdi Zajdi" on the *frula*, for example, where he demonstrates his control of the technique for several minutes.

but not prescriptive of any specific behavioural skill, nor indeed complete – some other factors may emerge as distinct and in need of an account.

The reader should pause here. The functional specification of the brain is to serve the Sequential Imperative. It turns out this is complicated. There is the general conceptual approach, with the brain as a kind of time-machine which transforms atemporal "inner world" entities into temporally patterned "external world" activity (and *vice versa*). In addition there is the management of the delivery of the SI, and the complexities of actually making it all happen. Contextualization is an important part of this (and may depend upon exploitation of the *primus inter pares* idea introduced earlier). In particular, context will work to enhance the performance of the (de-) sequencer with incomplete entity specifications, thus rendering the notion of transformation less precise than might be thought necessary.

What we need to do now, to help build understanding of what is going on, is make the account more personal and experiential. We need to balance, or flesh out, the abstract model and engineering system style of account above. To shift the perspective away from mechanistic management of the Sequential Imperative and more towards the experience we can change our terminology. The experience of managing the SI is, I suggest, better captured in the word "attention". Although we used the term in the more abstract account, we now do so in a more colloquial style. The material will necessarily conflate some of the points teased apart above.

3(7).13.4.1 We when look at something, say a bird feeding on a bird-table, lines of text in a book, a painting in an art gallery.... we are usually unaware of the patterns of fixation with saccadic shifts from one to the next: rapid movement of the eyes from fixation to fixation, several times a second. These movements are not random. The visual processing system directs the saccades so that the fixation points – automatic attention, if you will – serve to support the visual system as the visual scene is interpreted. The effect of this, in terms of the general model of spine and half-planes, is to manage the incoming data at quite a fine scale of analysis. The control of the saccades is monitored through their effect on the data flow. Saccadic behaviour is thus a way of imposing (and utilising) sequentiality – onto or into the "visual world" (without the animal being aware of the saccades or the fine scale of attention). Clearly the distal stimuli may be time-varying because of motion, say, or changes in illumination, or whatever, and on various timescales. The proximal stimuli have to be caused to be time-varying when they are not varying in consequence of distal stimuli changes, else the necessary sequentiality in perception will not happen and visual input will not work. (See also the quotation from Neisser, 1967, given earlier – 1(1).4.1).

3(7).13.4.1.1 In addition to the saccadic manifestation of attention a person's visual attention may be "focussed" on a particular part of the environment – say the gait of a horse, or the facade of a building, or whatever, at the expense of other aspects of the environment (such as the number of people in view, and their clothing). Selectivity is thus operating but it is a process, not a state, as we recognize when we shift attention from one thing to another. Attention is thus the activity (behaviour) of continuously focussing on something (rather than another thing, at that moment), with its attendant patterns of specific fixation overlaying the saccadic fixations of which we are not aware (examination of the lintels over the windows in the building we are looking at, then the gaps between the windows, then their height, and so forth; take "a look" at the image used on the cover of this book). There is a richly layered quality to this attentional activity (but it is not hierarchical; it is multi-scale) – the viewer is looking at one building not the next; at one part of the building then another; that part is being examined in a sequence of glances at different details (on the basis of knowledge of other similar instances); the saccades are being delivered in support of this attention but separately controlled. All this attentional activity is itself behaviour, driven by the SI as the expression of the intention to examine the building, to "look at it" in the sense we do rather than as a camera does. And because it is behaviour, and thus sequencing in its own right, attention can be monitored via de-sequencing.

3(7).13.4.1.2 The half-planes (tiers) and spine functional model seems inadequate but it does serve well enough descriptively, even if we need constantly to be reminded that it cannot diagram on paper the many thousands of half-planes and processes we assume to be operating all the time. The functionality is complex in extent, if you like, but not in principle. As noted earlier the half-planes and spine model of the attentional process is the use of, or favouring of, some of the half-planes as *primus inter pares* in their domain (many domains, so many half-planes selected as representatives). The favoured half-planes may vary from attentional focus to attentional focus – but these foci can be quite enduring so the effect on processing can be pronounced. Within any cluster of half-planes (domain) the identification of the favoured half-plane may vary from task to task, as we noted before in respect of syllables (2(6).11.3.1). But there is an additional factor to be kept in mind when considering attention (and ultimately all (de-) sequencing). Attention spreads forward in time, as well as covering the present and the past. This is elaborated below (3(7).13.4.3), but first we can reflect a little on another aspect of the model.

3(7).13.4.2 In respect of the size of the brain, and of the organism served by the brain, the model with half-planes and the spine needs to be envisaged as having the "right" overall size for work to be done. How can the development of the model (brain) to be the "right" size also be envisaged? An example might be helpful. In the development of small computers – personal computers – the processors have increased in speed, and the amount of available memory has also increased, over the course of time from the early 1980s. Indeed the trend continues. These incremental changes led to changes in performance, and in perceived performance, but the sense of functionality did not change for quite a number of years. At a certain point in the development of these machines it became possible to manage the manipulation of full colour images in real-time – the user could move a physical object (joystick or mouse) and watch the image move around or be transformed in one way or another; no jerkiness, no latency – just apparently direct coupling of hand motion to depicted motion on the screen. This change was experienced as a step change in functionality but it reflected merely a trend in hardware refinements over a time – something which continues (driven in part by the games industry). My point is simple – from one perspective a change seems radical; from another it is merely a point on a graph of continual change.

So it seems likely, I am suggesting, that from the point of view of management of the Sequential Imperative evolution has wrought a succession of changes which in one sense look radically different, even though the functional specification of the brain has not changed. The incremental (species by species) evolutionary addition of half-planes (tiers) to the model, with appropriate linkage to BGP and the spine, is all that is required. (Of course, species differences may emerge which can be accounted for by changes in half-plane groupings, not simply increasing the number of them.) Whether or not more changes are to be expected is a matter for conjecture. We will consider this later, as well as the possibility that there have been some changes which did not work out well (I'm thinking here of *Homo neanderthalensis*). We might also note in passing that management of multi-tasking is also behaviour for which additional half-planes need to be added; we will return to this. It is important to recognize that the incremental changes must not be read as some sort of linear progression from simple organism to Man. The emphasis is on incremental not on "progress". Species are incrementally differentiated, but the differentiation is multi-factorial, so there is no linear progression. (See also 3(7).14.3.)

3(7).13.4.3 There is another point of detail which needs to be clear in the reader's mind, and some further discussion/ illustration will help with this. The duality of entity structure (GCP2;(ii)) ensures that attention can be envisaged as favouring some entities over others in the control of behaviour (sequencing) and in perception (de-sequencing). But, in addition, at any moment in time some entities will not be completely specified because some of their component entities "lie in the future" and are thus specifications of upcoming moments or events, or anticipation of incoming but not yet experienced data. What attention does is to select how the "future" is shaped – specific entities are selected to drive both the sequencing and the de-sequencing. Another way to envisage this is that the nature of the (de-) sequencing processes means that the mapping of entities onto the spine will at any moment be partially incomplete providing some "look-ahead" or anticipation (prediction, in essence) of yet to be (de-) sequenced entities (cf. Clark, 2013). Attention works to shape this "future" aspect of the SI as much as the current (and indeed the past) aspect, by specifying and selecting entities which matter more than others (*primus inter pares*). Note that none of this implies that the cognitive entities involved are sequenced in their rôle as precursors – merely that their specificational contribution to the actually sequenced behaviour, or to the incoming de-sequenced sensory input, may be incomplete at the moment in time being considered (and note the rôle of context, discussed above 3(7).13.4). Let us take a longer, but even more informal, look at this issue (not covering all the 10 factors identified earlier).

3(7).13.4.3.1 A man is walking down a street in late spring – or more precisely, on the pavement, which is wide, and adjacent to the busy roadway which has two lanes of traffic in each direction. There are plane-trees (perhaps *Platanus × acerifolia*) with new season's foliage, growing near the edge of the pavement away from the shops and apartment blocks (and with roots that disturb the flatness of the pavement). The pavement opposite is just as wide and similarly organized. The width of the pavement permits cafés to set out tables for those wanting a coffee or a light snack, under the shade of the trees which also mute the traffic noise a little. Florists set out their displays on the pavement as well, as do some shops selling fruit and vegetables, or books, or whatever.

3(7).13.4.3.2 The man is a creature of habit and is on his way at the usual time to a favourite café where he is a regular. The waiters anticipate his arrival by putting his favourite pastry (almond croissant) to warm through – it is fresh but they know he likes to have it warm. His route takes him past a florist and as he approaches the florist he notices (i.e. attends to) a couple of things. First, in the background, he can hear an approaching "emergency vehicle" of some sort coming up the road behind him. He has learnt that there are differences between the sounds made by ambulances, police cars with all the garish markings, and "unmarked cars" on their way to what must surely be important business (rather than lunch). He thinks the one he can hear is an ambulance, and he knows it will pass him so he need not turn round to look – his identification will be (dis-) confirmed in the fullness of time, after it passes him and he can look at it without effort. He is, in effect, marking a moment in the future for specific attention to a specific event – the passing of the vehicle.

3(7).13.4.3.3 The second thing he notices is a bee heading towards him at about head height, from the florist shop, and on its busy way somewhere else to forage, or maybe back to the hive. He notices the bee only because he has to swerve to avoid it (well, it would have swerved, but he moved first because he has better vision and saw it first). He is reminded that he once thought to stop and have a chat with the florist about the bees he sees quite regularly – to ask about the presence of hives in the city, or perhaps if it is known if the bees are from out of town. He thinks it would be good to sort that out because later in the day he is due to give a talk to his students on "bee communication" and the supposed "language" they are described as using (the "waggle dance"), about which he thinks there is much over-interpretation. A local quotidian illustration of bee behaviour might help bring the lecture to life. He turns into the florist's shop, and thus changes some of the patterns of attention he had projected into the future. He might realise that he is going to miss the opportunity to confirm his identification of the emergency vehicle, or not – we don't know that yet.

3(7).13.4.3.4 It transpires that a few hives are kept on some nearby roof-gardens and the bees forage quite widely in the area – visiting several florist shops and the many window boxes full of flowers at this time of year. He could even place an order for a jar of local honey if he so desired. Pleased with this information he continues on his way to the café noting as he walks along that the heave in the pavement, caused by the trees, seems particularly bad – walking is almost like being at sea, he reflects, but it can't have changed so maybe he just attended to it more than before. He arrives at the café "late" and is chided by the waiter. He replies that he is not "late" but that he stopped for a chat with the florist on the way. The waiter (who seems to be from central casting) smiles – "a charming lady" – but our lecturer merely responds that he was talking about bees. With another smile and "naturally" the waiter goes away to get the usual coffee and croissant. The lecturer finds a vacant chair and table on the pavement and sits down.

3(7).13.4.3.5 The regular pattern of his morning looks like it will continue to unfold without further distraction – except that he pauses with coffee-cup held half-way to his mouth and realises that in fact he never checked the identity of the emergency vehicle which must have passed him whilst he was conversing with the florist. Disliking unfinished business he asks the waiter as he pays and gets up to leave, and is pleased to learn his identification was indeed correct.

3(7).13.4.4 The scenario just described is actually (so we can suppose) captured on a sort of film – no "arty" camera work, instead a calm capturing of a few minutes in the life of a lecturer. However, the scene is recorded on rather unusual "film stock". Let us move back in time, in our imagination, and suppose that the filming is stopped at the moment he comes out of the florist's shop and turns to continue on his way to the café.

3(7).13.4.4.1 Each frame of the film that has already passed through the camera records "the past", as it were, from the perspective of the frozen moment. Now, let us imagine, in a flight of fancy, that the frames yet to run through the camera are not blank, but neither are they fully composed or "filled in". Which is to say, some aspects of the contents of the frames are sketched in (his walking to the café, especially the movement of the limbs with ballistic, rhythmic qualities), and some details are less well defined than others (those which are more remote in the future, so to speak, and those which might be dependent on perception, such as movement of the head, or adjustment of gait to suit the pavement irregularities). Some aspects of his visit to the café are sketched with partial detail (difficult to do on film, but stay with it). For example, the intention to sit down and have a coffee and pastry is clear; the precise seat and table much more contingent on the situation he finds when he gets there. His attention runs into the future, from the moment of resuming his walk to the café, and thus determines in part what is sketched in to the "unexposed" frames. In fact the "film" captures multiple "attentions" all stretching out into the future.

3(7).13.4.4.2 The situation being described is specifically not that of unexposed film being run through the camera as the events unfold and time moves on. Rather it is that the future is partially specified from the perspective of the moment in time we chose to "pause the camera". This becomes clearer if we recognize that cognitive entities being sequenced for output behaviour – walking, talking, etc., – inevitably provide partial specification of multi-tasked actions *yet to happen*. The complex structure of the entities being sequenced (arranged on half-planes around the spine if you like, in a really dizzying system of half-planes) is not organized so that entity "boundaries" align in time. The "film" could be stopped at any of the thousands of frames in the sequence, and it will always be true that some detail is already being sketched out in yet to be recorded frames. Attention, in this "cinematic" sketch, is simply the detail flagged as significant in the past, threaded through the present and sketched into the future; the entities which matter more. The rest gets filled in as the frames move closer to the moment of capture. In this account then, we recognize that attention, with all the multi-tasking implications, is as much about prediction as it is about avoiding tree roots. But note, attention is idiosyncratically managed (context dependent) and so, therefore, will be the predictive element in perception and behaviour, taking us beyond external contingencies and dependencies to the realm of internal contingencies and dependencies. We will revisit later the implied heterogeneity in perception and behaviour.

3(7).13.4.4.3 The asymmetry between perception and action is not quite as clear-cut as might appear at first (as is the asymmetry between vision and speech, say; we will return to this). Perception – the de-sequencing of entities – will also involve some forward projection or top-down contribution to entity specification (especially those the focus of attention), merely awaiting (sufficient) confirmation from the senses (error correction is managed this way). Thus the bi-directionality of the sequencing and de-sequencing "mechanism" (whatever that is in the brain; it certainly isn't a notebook with thousands of pages, or anything like the computer program Pantome) ensures that the moment of investigation – the "stopped frame" – is always going to be a somewhat fuzzy point between the completed past (along with some completely specified entities) and the yet to happen future. And this gives us yet another way of considering the complexities of attention and the management of the SI.

3(7).13.5 The foregoing account is actually a description of a notion not yet mentioned – the notion of "now" in the flow of time – and it deserves a little elaboration. Some reflection on the sketch above suggests that "now" is a rather fuzzy experience. There are two significant conceptions to be worked through. The first is that in reality even completed entities, perceptions especially, are subject to subsequent corrections (and recall the discussion in Chapter 6). Ambiguous sentences are famous examples (e.g. *the man saw the boy on the hill with a telescope* could involve telescope assistance or possession), and garden-path sentences also pose problems requiring re-analysis, or re-perception, or re-de-sequencing (e.g. *the horse raced past the barn fell*). Actions are less correctable in this sense, but interactions and dialogue can require ongoing repair of what was already completed. In this account the "now-ness of now" is fuzzy because it is actively being worked over. Clearly "now" can refer to a brief moment in time – the catching of a ball – and it seems this cannot be fuzzy. But maybe some fuzziness is recognizable even in such a simple case through a discussion about whether a friend's exclamation ("Good catch!") could have disturbed the success of the catch, or could not, because it happened during or after the catch (the perception of the alignment/ overlap is what matters). Within the terms of the model with half-planes and the spine, what we are considering in the possibly dropped catch is that the boundaries between entities will not be cleanly defined; at finer scales of analysis there will be fuzziness in the alignment of sub-component entity boundaries (and see 2(5).10.2.4.2). Such perceptual imprecision in alignments of different events will contribute to a sense of fuzziness at the shorter timescales we are considering here.

3(7).13.5.1 The second conception of "now" we need to think through is that fuzziness can derive from scale effects. Into the future there is a wide range of entity specifications lasting from seconds to days to years even, where partial specification of yet to be completed entities is provided by projecting values into the entities yet to be fully completed. Attention identifies the entities carrying the burden of projection into the future, but these entities can vary considerably in required temporal scale. The point about "now" is that it is a reflection of such incompleteness spanning both the past and the future. Consider something more enduring – the "now" of having a meal with friends. This will have started some time back – maybe an hour or so before the moment of scrutiny. And the meal is still to be completed, so "having a meal" is spread over a couple of hours and with it the sense of that meal as defining a "now" in the life of the diners. If one of the guests turns to another – say our lecturer friend from the café – and asks "what are you doing now?" the range of possible answers fleshes out just why the notion is so fuzzy.

3(7).13.5.1.1 One answer might be "chewing a piece of meat" which could be true and would refer to a brief moment of time as the present, the "now" of the question. But it would probably be thought really weird as an answer. The pragmatic (contextual) implications of such a statement would need to be thought through (did the lecturer just take possession of a new set of dentures (a fact known to the interlocutor) and thus is cracking an obscure joke about how effective the dentures are?). More probably such a response would be interpreted as a rude brush-off roughly equivalent to "none of your business" (see 1(3).7.4.5).

3(7).13.5.1.2 Another answer could be "having lunch with friends" which elects to recognize a different timescale of event as the "now" – the lunch-time duration. This too is open to various interpretations, ranging from weird to completely acceptable, depending on the understandings shared between the two speakers. I leave it as an exercise for the reader to conjure up the various scenarios which could contextualize the exchange with this particular response. It is interesting that we can do this "conjuring up" of scenarios relatively effortlessly.

3(7).13.5.1.3 A different answer to the lunch-time query might be "working on a lecture on bee communication". This too could be interpreted as evasive, but not offensively so. The interlocutor could pursue this topic to learn what is new or exciting about it. Perhaps the lecturer is now doubtful about the use of the word "communication". The interlocutor could thus humour the lecturer before asking "but what else are you doing?" which hints at a longer timescale than mere lecture preparation. Note that the lecturer's initial response has an ill-defined sense of "now" because he does not reveal when he started working on the lecture. Unless he also lets the interlocutor know that the lecture is to be given that afternoon (which we knew, in this story-line) the future extent of the "now" remains undefined and acceptably so (the interlocutor might want to ask about that, but need not). Our lecturer's subsequent response to further probing might reveal a willingness to consider longer time-scales. "This year I'm working with Fred on a couple of new modules" defines "now" rather clearly, but could still be construed as evasive if the interlocutor knows the lecturer to be a keen researcher with some big projects completed and several more in negotiation. The phrase "doing now" could be interpreted to refer to a new job (depending on when they last met). At this scale of analysis the "now-ness of now" reaches back perhaps a year or more into the past, and forward perhaps a few years.

3(7).13.5.2 Intriguingly, all these senses of "now" are available to the participants in the conversation, and with sufficient clarity that the pragmatics of the situation pose a distributed cognition type of problem for resolution in the dialogue. Whatever answer is given to the initial question is known to both parties to be a selection from a range of possible answers with different "now-spans" in time. This is a rather more explicit notion of attention, selecting entities to pick out portions of time ranging back into the past and potentially some way forward into the future, where the entities provide the sense of "now".

3(7).13.6 The above excursion into fiction serves to illustrate the multiplicity of co-existing overlapping entities of different durations in sequenced activity as behaviour, action or event to be perceived. At any one moment in time the brain serves to support the Sequential Imperative with many cognitive entities in various stages of specificational completeness. Attention will serve to manage which entities matter most in the processing – there will be many of these and the management of the processing does not require that the *primus inter pares* entities identified for the processing remain the same indefinitely. Indeed, attention is as much about how best to control the SI as it is about actually providing control. Attention in the management of the SI is about the management of the "now-spans" that contextualize the processing for both sequencing and de-sequencing. Note that the range of "now-spans" is simultaneously experienced in one's introspection of "now" – that is how the discourse pragmatics work with the range of answers given in the dialogue sketched out above. Thus one's sense of "now" is not a specific portion of time, but a set of different overlapping portions of time experienced in a synthesis of both time and timelessness; and also a synthesis between experience and prediction. It is as if we experience the multi-layering of attention at different scales as a sort of atemporal and thus internal representation of the flow of time. We call this consciousness – a topic to which we will return – but note here that it has no functional purpose, it is the experience of attention and thus of the function of serving the SI.

3(7).13.6.1 The attentional complexities just explored are those we might expect in humans – but we can generalize away from lunch with friends to a scenario where a creature's brain is just large enough (relatively; this will be addressed later) for the management of attention itself to be behaviour, and this becomes the limitingly small case of managed multi-tasking. We can consider this limit to be the case where attention is being managed as a task along with one other task. The suggestion is that if there are many such other tasks we are indeed considering *Homo sapiens sapiens*. If there are fewer such tasks, and multi-tasking is limited, we may be considering other primates (chimps cracking nuts), birds (building a nest), and maybe some other mammals (this is yet to be determined).

3(7).13.6.2 The selection and deployment of entities at various scales in the Sequential Imperative processing is what constitutes behaviour – usually observable when the finest scale entities are lined up in sequence as "output", with the larger precursor entities inferred from the sequence observed (cf. syntactic structures). Attention is no different – involving the selection of entities at various scales and sometimes with observable output (e.g. eye-gaze, tool usage). Attention is a form of behaviour although its output is not always obvious or even required. But, and this is intriguing, when observable attentional behaviour is desirable we may well end up using language to achieve and manage that observability. Furthermore, if language is considered another attentional mechanism then the evolution of the brain for language is effectively just part of the evolution of the brain to deal with more multi-tasking. Indeed, perhaps the better way to think about this is that language emerged because multi-tasking was becoming more complex and needed complex attentional behaviour (i.e. language) to manage it. The management of the SI, in this scenario, becomes complex enough for language behaviour to look necessary rather than accidental. The language behaviour will be communication, not inner speech. (See also 3(7).13.8, and the next chapter.)

3(7).13.6.3 Affect is an attentional mechanism (GCP6) in the sense that emotional states and behaviour can drive the attentional processes (flight, hunger, bonding…) by providing some of the larger entities required to "entrain" the Sequential Imperative at finer scales and thereby direct behaviour in ways similar to "normal" attentional behaviour. It is the case that the cognitive processes producing normal attentional behaviour can also evoke affect as a mechanism, but also likely that affect is invoked as a form of attention in non-cognitive modes (hormonal, recognition of emotion in other beings, empathy, etc.). Affect seems to be a necessarily observable form of attention and SI management (so, for example, crying serves the purpose of providing an observable behaviour indicative of affect but also susceptible to the demands of the Sequential Imperative).[5]

3(7).13.7 In sum, therefore, the Sequential Imperative delivers behaviour but is managed by behaviour at a different scale. Figure 5.7 is not an illustration of lots of different spines with half-planes radiating out to the cognitive hinterland. Rather it is an illustration of the huge variation in scale to which the diagrams can apply. We earlier used the term "now-span" to indicate the temporal extent of application of some of the entities in the sequencing. An intention to make and serve a meal provides the longer now-span which contextualises and drives the finer scales of entity enabling the sequencing which actually produces observable behaviour (and the food to eat). Affect can work similarly (GCP6). The meal making now-span is merely a brief moment in the longer now-span of attending a cookery course lasting two weeks; itself a brief moment in the reconfiguration of lifestyle skills in a person's retirement; and etc.

3(7).13.8 GCP4 covers learning. Attention, as behaviour, has to be learned along with everything else that makes it possible for the brain to serve the Sequential Imperative, and this includes affect as an attentional mechanism. But there is another intriguing topic that needs to be included here. Learning serves the

5 In the context of an evolutionary account of the development of the (de-) sequencing engine it is possible that affect became necessary for control of attention in homeostasis, just as language is necessary to manage attention in much more complex brains with rich multi-tasking opportunities.

SI – and that must, therefore, include learning language(s). As we noted above, language might better be understood as an attentional mechanism. Interestingly, mother-child interactions in a language learning scenario are often about attention, and certainly require attention to work. Maybe language acquisition is as much about learning how to manage attention as anything else?

3(7).13.9 What we have seen in this Chapter so far is that the rather mechanistic or engineering style of account of SI management as a systems problem (how could it be made to work?) can be complemented by a more experiential account using the term "attention" which is readily given an appropriate subjective interpretation. In addition, the engineering issues have been set alongside the subjective notion of "now".

3(7).13.9.1 The topic of "fuzziness" provides a nice link between the two perspectives. The alignment of entity boundaries is fuzzy – for example the production of a phone will require various distinctive features to be "switched on or off" notionally in synchrony. In reality this doesn't happen, but categorical perception smoothes out the discrepancies. Segment boundary identification is a long-standing issue in Artificial Intelligence research and surfaces here as an "engineering style" issue in discussion of the delivery of the Sequential Imperative. The use of the term "fuzzy" in discussion of segmentation boundaries can be contrasted with the use in the discussion of "fuzzy now" where scale effects predominate and incomplete entities (which may be about long periods of time) all simultaneously contribute to the sense of "now". In this account "now" is not a systems issue, but a perceptual and experiential one. Subjectivity complements the engineering. And, as has been suggested already, we can see a way to developing a notion of consciousness as the experience of attention. Yet another bonus is the possibility that the emergence of language could be understood in terms of the emergence of complex attentional behaviour in beings capable of multi-tasking activity. We will encounter these two points in the final two chapters.

3(7).14 There is one additional concern about how attention works which we need to consider before moving onto the next chapter with its focus on evolution. We have introduced some terminology in the material above: clusters of entities in domains are variably represented in other clusters by specific entities (the *primus inter pares* notion) which can, therefore, be thought of as representing those domains, and so on at different scales. We can envisage the representational entities as attentional foci – different "now-spans" when we consider the actuality of behaviour rather than its past or its future. There is, however, a question. How many such attentional foci could a creature's brain require for delivery of the Sequential Imperative? We return again then to the issue raised earlier – that of scale (cf. 3(7).13.1.2).

3(7).14.1 Creatures with brains vary considerably in size, as indeed do their brains. There are several ways to summarize the inter-relationship. Overall, as body size increases so does brain size. However, smaller animals have relatively larger brains, as a proportion of the overall mass, than do larger animals. The details of the relationship and its characterisation are not relevant at this point (see Chapter 1, FN4) – see the discussion in Martin (1998), and Schoenemann (2006) which goes on to look at evolutionary changes in the human brain as well (3(8).15.6.1.1).

3(7).14.2 Our interest here is that as the functionality of the brain is to serve the Sequential Imperative, and as differences in the expression or manifestation of the SI amount to species differentiation, we expect to see variations between brains caused by body size, sensory variations, articulator differences, patterns of locomotion and feeding, and so forth. However, one important source of differentiation is attentional behaviour. This too is part of the SI species differentiation, and will be reflected in the detail of differences between brains. Of particular note is the necessary link between increasing complexity of attentional behaviour (and with it brain size and structural complexity) on the one hand, and articulatory complexity and multi-tasking behaviour on the other. An example might be the development of bi-pedal locomotion, which frees two limbs/ hands for other uses. But unless the delivery of the SI, with appropriate attentional foci, yields "control" of the freed articulators for tasks other than those readily accomplished by one front limb in animals with four footed locomotion (scratching for food with a front foot/ paw/ etc.), then the potential for multi-tasking is not going to be realised. Evolution, in this account, requires something which looks like "co-adaptation" to deliver multiple interacting changes. The brain serves the SI but must evolve with any changes in the articulators/ sensory organs, else the SI cannot be "fully" served.

3(7).14.3 The requirement for "co-adaptation" must be present *within* all beings capable of evolutionary development. This requires some deliberation. The use of terms like "co-adaptation" and "co-evolution" is normally reserved for describing situations in which two species evolve or adapt together in some sort of mutual benefit scenario. My concern here is for co-adaptation within an organism. This does not seem to be widely discussed. The issue is readily illustrated: if, say, a primate evolves an opposable thumb then one can refer to this emerging in the usual manner (sexual reproduction is taken as the main means of introducing hereditable genetic variation; progeny are "over-supplied";

progeny vary in survivability). But this obscures the account of what is actually changing in the organism. The opposable thumb requires an adjustment to the carpometacarpal joint, to the musculature, to the innervation of the muscles, and to the neurological configuration providing for the control of the thumb. These have to co-adapt, in my use of the term. However it is proposed (in the shelf-load of relevant books) that this co-adaptation is possible in any organism, it underpins evolution in the way an organism's brain delivers the Sequential Imperative for that organism. For our purposes here it seems easiest to say that speciation is the result of evolution yielding differentiation in the details of delivery of the SI. And of course the SI is concerned with perception as well as action, so evolution in sensory apparatus requires the same sort of co-adaption in an organism. The co-adaptation is necessary, of course, otherwise evolution would not happen.[6] It is not our concern here to explain how it is achieved, merely to accept that it is must happen. Note, incidentally, a recent report that *Homo neanderthalensis* evolved larger eyes and with them more of the brain was devoted to vision (Pearce et al., 2013). This could have imposed a constraint on evolution of more complex attentional mechanisms – vision in crepuscular gloom was enhanced but maybe the potential for multi-tasking diminished? Which is to say, co-adaptation does not imply long term success for an organism; it does not filter out "poor" adaptation.

3(7).14.4 In the next chapter we turn our attention to a discussion of language – the "evolution of language", as some people describe it. I hope to show that this is a mistaken way to think about how humans come to have language behaviour. We will build on the above outline account of attention and its evolution, including the suggestion that language is merely an attentional mechanism – it is primarily about serving the Sequential Imperative, not about being governed by the SI (although it is that too). We will also revisit the idea that consciousness is the experience of attention. Specifically, we will look at the "evolution of language" from the perspective of Universal Grammar, alongside a couple of unconventional accounts of the more usual "evolutionary" story – accounts which focus on the syllable.

6 We will consider the details in the next two chapters, but for now we can note a few complexities. Firstly, the details of the functional model (sensorimotor "interface" with the outside world; dizzying number of half-planes; background processors) suggest that evolution in three domains has to be reconciled – but then again, it is just a model. Secondly, synchronized co-adaption in two domains is not required if the first change is useful in its own right. The second change then exploits the first, in a process known as exaptation. One is reminded of Gould and Lewontin's discussion of spandrels (1979).

3(7).14.5 The conceptual core of the book has been covered – the interpretation of the Sequential Imperative in terms derived from Part 2, and for reasons implied by the linguistic material used there. The inescapable Sequential Imperative carries with it a concomitant requirement that it has to be managed. Disruption in the management will bring unusual behaviour. But, and this is intriguing, successful management in ever more complex brains may require the emergence of language as an attentional mechanism for the management of multi-tasking. Animal calls, in the general case, already show us that "communication" is actually about attention and its management, and so it should come as no real surprise that in humans language behaviour can be characterised as being about attention. In the remaining two chapters we consider implications for "evolution of language" and for notions of mind. Before getting to that point, however, the pan-species aspects of SI management need to be noted. Clearly, some of the factors listed earlier may be specific to humans, some are common to any creature with a brain, and some may be a matter of degree (see 3(7).13.4). Multi-tasking and attention as behaviour may be specifically human – but even that isn't compellingly obvious. Managing the SI has to be done, regardless of brain size, and is not trivial.

3(7).14.5.1 On evolution in general, our stance is that this must be a sufficiently complex process that co-adaptation, in my sense (where muscles and sensors and brain within a species co-evolve), always delivers for the organism a brain that can serve the Sequential Imperative. Differentiation in the serving of the SI is speciation, on this account. But note a tension here between differentiation yielding speciation and differences between members of a species yielding heterogeneity in the delivery of the SI. The *primus inter pares* conception in the organization of half-planes and domains (see Chapter 5) will yield heterogeneity in (de-) sequencing and thus in the achievement of skills. Cognitive (and thus linguistic, in the case of humans) homogeneity within a species is not to be expected. We will return to this when considering language and thought.

3(7).14.5.2 Further project work identifiable here concerns patterns of disruption in the control of (de-) sequencing. For example, it may be that behavioural patterns identified as "deviant" in some way can be traced to, or aligned with, aspects of the (mal-) functioning of the (de-) sequencer. In addition, the significance of learning in relation both to management of the Sequential Imperative and to the background "off-line" domain where memories are stored and goals formulated, etc., needs deeper investigation.

CHAPTER 8

Issues in Evolution and Language

3(8).15. In this chapter we consider complex "behaviours" such as speech, signed language and gesture, in humans, and we look at how these topics provide a general insight into the evolution of behaviour in creatures with brains. There are some terminological niceties which need to be clarified before we get into the detail. In 1(2).5 we noted the third General Cognitive Principle:

GCP3 Behaviour is sequencing; Perception is de-sequencing.

Behaviour, in the sense meant here, covers "activity" or "action" or "doing things" or "behaving" in an observable sense. As we will see, there are some additional refinements and elaborations. But what matters here is what is not meant. The sense of behaviour we have here does not submit to pluralization. "Behaviours" such as *walking, barking, talking, nest-building, courtship behaviour*, etc. are poorly and/or incompletely defined terms, used loosely in descriptions of aspects of behaviour in the sense we do mean. There is a tendency to use these summary terms as replacements for the detailed account of behaviour we should be considering. Thus, to return to our opening sentence – we will be considering speech as behaviour, not as "a behaviour", and the work to be done in this chapter is in part concerned with showing how the "loose talk" terminology can be set aside, indeed must be set aside. One particular misconception we need to sweep away is that "a behaviour" evolves – it doesn't; the behavioural apparatus evolves (and is exposed to behaviour as well as managing behaviour itself). Visible behaviour may change as a species evolves; "the behaviour" (nest-building, whatever) does not evolve. The misconception bedevils discussion of the emergence of language in evolution because of the assumption that language evolved from some other behaviour (gesture, tool usage, proto-language…). The first part of this chapter clears away the misconception. Subsequently we move on to consider how our phonology-based model of the Sequential Imperative provides a general insight into the emergence of behaviour in creatures with brains.

3(8).15.1 The care with which we use the term behaviour matters when we turn to consider evolution. It is behaviour which *changes*, because the potential for sequencing and de-sequencing *evolves* (think of the number and clustering of half-planes in the model discussed in 2(5).10.3.6, see Figure 5.7). "A behaviour"

© KONINKLIJKE BRILL NV, LEIDEN, 2017 | DOI 10.1163/9789004342996_013

such as language does not evolve in the sense of interest to us here. To be sure, as Jackendoff (2002) reminds us, there is a sense in which languages (plural) "evolve" culturally, but he argues that they are already languages – and it is how we got that far that is of interest here (as in Jackendoff, op. cit., MacNeilage, 2008, and Carstairs-McCarthy, 1999). We will revisit that assumption a little later. I will start below with Jackendoff's account of the "evolution of language", returning later to MacNeilage (2008) and Carstairs-McCarthy (1999). We will also look at Berwick (1998) and Berwick and Chomsky (2016), before moving back to the main focus – the evolution of the behavioural engine. This coverage makes no effort to contrast different approaches as one might when looking for the better account. They are all, but differently, inadequate.

3(8).15.2 We need to focus on language because (a) the model developed earlier is built upon ideas derived from phonology (and we will be going back to this topic); and (b) the part of behaviour we call language is the most studied of any aspect of any animal's behaviour, yet apparently resistant to evolutionary accounts devoid of special pleading. We focus here on clearing the background confusion (as I see it) concerning the "evolution of language" before considering the evolutionary account of behaviour generally.[1] My aim is to leave the reader comfortable with the idea that it is loose talk to speak of the "evolution of language". Notice, incidentally, that "communication" or "language" is not given its own General Cognitive Principle. This is simply because such behaviour is behaviour, not a matter of principle; it is learned as is other behaviour, and it serves the Sequential Imperative, with which it also has to conform, as does all behaviour. If it were to be given a GCP we could envisage working that up from the notion that language emerged to serve the management of the SI, that it is indeed an attentional mechanism alongside affect (but my problem would then be that language is not general enough across all species with brains to be given a GCP within the discussion of attention). Furthermore, I have chosen not to work up a GCP for communication because I think the issue is open to debate. Language may have emerged to serve the SI by managing attention in multi-tasking, but arguably does more than that now – or does it? Leave that question to one side for the moment.

3(8).15.2.1 Jackendoff (2002) discusses the problem of the "evolution of language" (as he wishes to see it) as being that of the elucidation of the sequence

1 When I use terms such as "evolutionary account of behaviour" I am not contradicting my earlier point about behaviour not evolving. The reference is shorthand for the evolution of the behavioural engine. The "loose talk" is never far from the ink!

of phases (essentially of increasing linguistic complexity) by which language as we know it might have emerged. That some have proposed rather startlingly "all or nothing" accounts of the end-state, with no routes possible to get there (leading, inevitably, to innatist accounts), can be taken as the starting point in speculation – they point the way for reasoning in the search for a sequence of "steps" by which a species can end up chattering away as we do, about the weather, the opera, and the latest images from the surface of Mars. Jackendoff (op. cit.) is clear on this, and aligns himself with Bickerton (1990, but see also Bickerton, 1998) and the notion of a "proto-language" (the model at this stage is a 2 step model). He goes on to elaborate a more complex account of the proto-language, one involving symbols – initially in small numbers, and then in large sets with syntax to constrain the ordering. He points out that large sets of symbols bring with them issues of discriminability in production and perception, irrespective of ordering problems. This in turn leads to accounts involving componentiality within symbols (combinations of segments; syllables and phonology), and then to the organization of symbols into strings. Simple concatenation is followed by syntax.

3(8).15.2.2 Jackendoff (op. cit.) attempts perhaps a middle way between the strongly innatist view of language as the outcome of a saltation in humankind's genetic endowment, on the one hand, and on the other hand a view of linguistic capability as arising as the consequence of general evolutionary diversification of capabilities in the species known as *Homo sapiens*. In the first case the requirement is that one or more big genetic changes occurred fortuitously, and even coincidentally, and that the resultant beings knew what to do with them, so to speak (we will return to this). In the second case the evolutionary changes were exploited – as they became available, step by step, over perhaps quite long periods of time – for the emergence eventually of linguistic communication in all its complex richness and multi-layered activity.

3(8).15.2.2.1 Jackendoff's middle way is perhaps the consequence of an unwillingness to relinquish the elegance and power of the theoretical construct of Universal Grammar, with its requirement for all or nothing emergence, whilst also recognizing the problem that it can't really be delivered by evolution. Yet at the same time he sees a problem with the gradualist account of the emergence of language-using capability in that it pays no heed to the linguist's understanding of the complexity of language. We really do need to consider how many muscles are required to produce syllables in sequence, and how many different scales of description are required of the linguistic material – all needing to be learned and managed for perception and production of speech. But, if we embrace his view that a few stages in development can take us from non-language to the full range of linguistic capability, following his development of Bickerton's notion of proto-language (see Jakendoff 2002:235ff), then in point of fact we have made no progress. Each of the steps proposed by Jackendoff might make sense linguistically, and seem "simple", but the requirement is for a series of genetic changes each being little short of the saltational miracle required by the purely innatist account.

3(8).15.2.2.2 To be fair, by the end of his discussion Jackendoff appears nearly ready to abandon the UG perspective altogether, referring to the "evolution of language" as "deeply incremental, adding more and more little tricks to the cognitive repertoire available to the child faced with acquiring a language". By this point he is suggesting that UG is "built out of layered subcomponents", and that "grammar is not a single unified system, but a collection of simpler systems". However, the problem left for the reader is that although he has provided an intriguing account of how the notion of proto-language can be elaborated, and of how a sequence of steps can be envisaged which add elaboration and complexity to "linguistic" structures, bit by bit – in complete contrast with an innatist "all or nothing" account – he has failed to tell any sort of story about how the account is evolutionary.

3(8).15.3 MacNeilage (2008) and Carstairs-McCarthy (1999) both see the syllable as key to understanding the origins of speech and language. We will consider their accounts in more detail later, but here it is worth noting that although they attempt to circumvent the miraculous in genetic change, there must be a concern, especially with the work of Carstairs-McCarthy (op. cit.), that somehow the task facing the scientist is to provide an account which features a succession of "behaviours" each of which is more complex than the previous. The similarity with Jackendoff's approach (op. cit.) is clear, although the sequence of steps to be gone through is different. The difficulty with MacNeilage's account (op. cit.) is that without an accompanying story about linguistic structures (which could be the account offered by Carstairs-McCarthy, but MacNeilage passes on the opportunity to invoke this) the account of the origin of discriminable sounds, systematically organized, doesn't lead anywhere. The reader is left with – where/ how do we get from there/ then (syllables) to here/ now (linguistic communication)? It should be noted that the problem, from within any of the perspectives taken by Bickerton (1990), Jackendoff (2002) and Carstairs-McCarthy (1999), is made much more difficult by the existence of so many different or "competing" linguistic theories. Just as the plethora of linguistic theories is problematic for the understanding of human behaviour with language, so it is problematic for understanding how we got to have languages.

3(8).15.3.1 The problem is that in principle there is an indefinite number of possible accounts with accompanying stories suggestive of reasons why one or more steps take us from "the behaviour X" inevitably to "the behaviour Y", which is "closer" in some undefined sense to where we think we want to be, in theoretical terms. The goal, of course, varies according to the degree to which one considers oneself bound by the thinking behind Universal Grammar, or any other linguistic theory. This is a bit of a mess, it has to be said. But the situation is made worse by the fact that evolutionary links, or changes, are sought in the behavioural outputs of the beings/ humans rather than in their behavioural potential.

3(8).15.3.2 Consider a parallel argument from a completely different domain – tool usage. The conventional, "old-fashioned", screw in carpentry came with a single recessed linear slot in the top or head, into which the user positioned a screw-driver blade. The user rotated the handle of the screw-driver, and thereby the blade imparted force to the screw, and the screw turned with it. More recently the slot-head screw has given way to one or another form of "cross-head" screw in which the business end of the screw-driver is a pointed cross-shaped sort of blade which is inserted into the recessed cross-shaped depression in the head of the screw. The cross-head screw-driver is much less prone to slip out of the screw than the conventional blade and thus the driving of the screw is more reliable. This permits the use of motorised screw-drivers to provide the rotation without fear of the driver disengaging from the screw. It is entirely nonsensical to say that the slot-head screw evolved into the cross-head screw. The underlying behaviour is the same, but in the case of the cross-head screw more reliable. Talk of evolution in the design of screws is loose talk. Talk of "evolution of language" is loose talk in the same way – the focus is on the superficial "observables" and not on the underlying capability. Well, actually the observable behaviour is not superficial in the sense "easy" or "insubstantial", as we have noted earlier. The point is that the complexities of linguistic behaviour imply complexity in the behavioural capability, and that complexity and its evolution are difficult to describe, as we see in this book.

3(8).15.3.3 Or perhaps – think of handedness in human behaviour. This is known to be indicative of some aspects of hemispheric specialisation of function in the brain. It is not sensible to speak of the evolution of handedness, with perhaps some stories about the need to use hands differently, and then some reason for favouring one hand over the other, perhaps linked to ready defence of the portion of the chest wall covering the heart, and etc. Obvious nonsense. The brain has evolved patterns of hemispheric specialisation for reasons yet to be determined, and one side-effect is handedness in observable behaviour.

3(8).15.4 The reader may, at this point, be unconvinced – but we are not short of additional argumentation from elsewhere in human experience with language. There is a totally different way to see the difficulty with talk about "evolution of language". The assumption made is that the end-state is well defined within a homogeneous speaker-hearer community. But this is not realistic. The force of this argument comes from two different directions.

3(8).15.4.1 The first issue to be considered is that of homogeneity in the speaker-hearer community of users of a language; let us call it ZAZ. It might be thought that users of ZAZ must share some essential understanding of ZAZ. It is not at all clear that they do, or indeed that they need to. What is required is that "enough" is shared for language use (and user pairs, so to speak, can differ in what they share). This includes techniques for the exploitation of context, as well as the exploitation of the pragmatics surrounding discourse repairs and semantic negotiation. Such techniques go a long way to circumvent the problems caused by the fact that it makes sense to consider users of ZAZ to be managing the fact of their idiolectal approach to language. In such a model the sense is that ZAZ is "a language" which exists simply as a group of idiolects with sufficient equivalence or overlap to permit the negotiation of more or less successful communication.

Notice that this approach contrasts strongly with the requirements of UG. The UG model has it that the underpinning competence for a language user is the same for all (first language) users of that language. Homogeneity is exceptionless. The triggering process (discussed earlier (3(7).12.1.2)) must work with input derived from performance grade processing and the possibilities for errors therein. But there is a problem additional to the one mentioned earlier. Competence underpins performance but users may make errors of execution, or input processing, yielding data which are less than perfect. Triggering is thus a performance issue and not itself governed by competence, yet it needs to serve the requirements of competence and the exceptionless homogeneity entailed therein. The competence of a language user is inscrutable – grammaticality judgements are performances; any other way we have of working with our competence is a performance. There is no requirement on language users (nor any way in reality) for delivery of exceptionless homogeneity of performance in the community. There is, in short, no evidence for exceptionless homogeneity of linguistic competence. There is a need for such competence according to the theoretical account of language in UG, but that doesn't make it happen.

3(8).15.4.2 Differences between groups of individuals are necessary in models of language using communities because otherwise language change could not happen. Returning to Jackendoff's point (2002) about languages "evolving" culturally – this concept can only be given meaning if one considers languages to be imperfectly or incompletely expressed in the same way, by a sufficient number of users, for changes to become manifest and to be propagated through a community and to new learners of any such language. The fact of language change undermines the notion of homogeneity of linguistic knowledge in the user community. Notice also that in some environments homogeneity is known to be absent – users of a pidgin "language" provide a linguistic environment for their children to learn/ create what becomes a creole. This is sometimes taken as evidence for the existence of innate linguistic abilities – required apparently in order to transform the disorganized input into a true language. It is not obvious that this account is credible.

3(8).15.4.3 Differences between individuals using "a language" are also to be expected, because the life experiences of individuals are idiosyncratic. Individuals have idiolects because indeed their whole cognitive world is individual. It makes no sense to assume that an individual's knowledge of language ZAZ is perfectly shared with other users of ZAZ when virtually nothing else is, or ever can be, shared this way.[2] The innatist account would have it otherwise, of course, because the genetic endowment provides just the mechanism required to underpin invariance and homogeneity. One difficulty is that the endowment necessarily requires configuration for the language community (for ZAZ as opposed, say, to AZA). But the configuration can only be managed on the basis of idiosyncratic processing of possibly relevant incoming data in the child's experience of the world, to yield the linguistically relevant (and specified) material required to engage with the innate endowment to set, perhaps, a parameter for headedness in phrases, or whatever. As noted earlier (3(7).12.1.2), precisely to the extent that this can be made to work, the endowment ceases to be required. Notice that this is not good news for UG in respect of just one child, but the problem is much more complicated if one considers a whole generation of children learning ZAZ, say, but with idiosyncratic experiences of ZAZ and of the rest of the world. Homogeneity is simply not available. Whilst it might be argued that UG makes the heterogeneity unproblematic, by providing homogeneity via the genes (one must suppose), if UG turns out to be unsustainable then homogeneity must be considered irrelevant, not just unavailable.

2 Consider the work we have to do to ensure that two scientists share the "same knowledge" of one or another theory or experimental outcome – or think they do. Similarly, children learning arithmetic or algebra struggle to have the same understandings as their classmates.

3(8).15.4.4 Probably the most common linguistic condition for a human being is for that person to be the user of two or more languages, acquired simultaneously from birth, and in the same manner, generally speaking, wherever they are. In some cases – perhaps most – the languages achieved by the child are related in some sort of language family or group, but this is not necessary (as in a child learning Spanish and Basque together, in a bilingual family setting). Indeed the very notion of language family can be challenged, especially in respect of the possible rôle of such families in human history (Trask, 1999).

Languages are not readily fitted into statements about universal properties of language – languages don't conform perfectly to one or another structural account (see Chapter 4). But, for those who wish to speak of "the evolution of language" this raises significant problems. It could be argued that language evolved only once, and got through to the fully fledged model without deviation or separation into a multiplicity of languages (which then emerged somehow, for whatever reason, in a riot of differentiation). But, this seems really implausible. So notions such as those offered by Jackendoff (2002) or Bickerton (1990) concerning "proto-language" and various incremental stages, therefore stumble on the fact of variation between the communities of users, and this would have to have been present in the "evolution of language" phase as language progressed through the proposed stages. There might be many "proto-languages" not just one, and many languages might emerge, not just one, and all in different population groups. But then, if language "evolved" more than once there is no reason to suppose that the various forms of UG would be compatible, so an argument for a single UG in all humans is an argument for the emergence of language just once, and from that original proto-language all others evolved (including signed languages) but conformant with UG which itself is assumed (without argument) never to change after the initial genetic change (80,000 years ago, see next paragraph). (See also Nichols, 1998).

3(8).15.4.5 Berwick (1998) discusses Chomsky's Minimalist program and the linguistic operation "Merge" in relation to "the evolution of language". Berwick and Chomsky (2016) view the saltational miracle required as a "minor biological change" that provided the linguistically required process, namely the operation called "Merge". This happened in a small group of hominids in east Africa, they suggest, more than 80,000 years ago. This enabled these hominids to have a "language of thought" but without speech (we might think of this as a rather abbreviated or incomplete form of inner speech), and then eventually speech itself. They write: "At some later stage, the internal language of thought was connected to the sensorimotor system, a complex task that can be solved in many different ways at different times". One difficulty is that by arguing for the primacy of language of thought, in a group, there is no reason to suppose (and no way to ensure) that the hominids concerned would have the same language of thought and thus the same spoken language when they got around to solving the complex task of connecting the sensorimotor system. The somewhat hand-waving approach to the sensorimotor realization of speech seems casual in the light of what we know about the complexities of making speech work (see Chapters 4 and 5). In this book there is no sense to be made of the notion of "language of thought" and our sense of internal/ inner speech is not considered to be evidence for a linguistic mode of thinking, (see the discussion in the next chapter, 3(9).19.6.7/8).[3] It therefore makes no sense here to consider the emergence of any internal "language of thought" as coming before the use of language for communication.

3(8).15.5 The outcome of this brief account is that it is probably impossible to flesh out any sensible or plausible concept of "the evolution of language". Language, and indeed languages, emerged as the consequence of evolution in the behavioural engine itself, the brain (eventually of *Homo sapiens sapiens*), coupled to the sensors and articulators required. This was not a single genetic change yielding UG, but one or more changes in the delivery of the Sequential Imperative and thereby the gradual emergence of the possibility of language behaviour. We turn next to consider this in some detail, but before we do we need to note that one of the drivers for interest in UG is the obvious fact that only humans have language. Whatever alternative account we develop for the evolution of the possibility of language has to come with species specificity. But then again, in the functional model we are working with here the point is actually made the other way around. Human species specificity expresses the Sequential Imperative in terms of particular management of multi-tasking complexity in the SI which happens to potentiate language, and then indeed rely upon it. We don't need a story about the human uniqueness of language which is isolated from other human uniquenesses.

3(8).15.5.1 In the previous chapter the point was made that evolution, in the general sense, must involve co-adaptation in both the brain and the "peripheral apparatus" of sensors and articulators. Further, it was suggested that expression of the Sequential Imperative will necessarily be different in different species – evolution means evolution in the way the SI is served by the combination of brain, sensors and articulators and consequently in the behaviour

3 Ian Tattersall, "At the Birth of Language", reviews Berwick and Chomsky (2016) in The New York Review of Books, 63(13), August 18, 2016.

available to an organism. It was also suggested that a correlate of this development in the expression of the SI is that attention evolves as do species: evolution in the delivery of the SI means evolution in attentional options. This is what is meant by saying that the behavioural engine evolves, rather than any specific style or range of behaviour such as we might recognize as language (including natural gesture and signed languages).

3(8).15.5.2 Let us take a step further with this line of thinking. Consider again the possibility that language is just attentional behaviour; indeed that language fits with other vocalisations in other species in being attentional behaviour (see 3(7).13.8). This is a cognitive perspective on the emergence in the animal kingdom of what we call communication. From this perspective what is it that needs explanation about the difference between human languages and other behaviour in other species? The academic discipline of Linguistics tells us, as we have seen, that human language behaviour is extremely complex – from control of voice onset time through to the simultaneous utterance of several propositions, and not forgetting that signed languages use a different articulatory medium altogether (and there is a further surprise to come on this, as we will see shortly).

3(8).15.5.2.1 Given that attention can also be behaviour (when not automatic), and that complex attentional demands in a multi-tasking environment require attention to be attended to (it is merely another domain within which the requirement for attentional foci has to be managed), the complexity of communication structures reflects the complexity of attentional behaviour. As the overall complexity of attentional behaviour increases then the possible complexity in communication behaviour will also increase. This is not language evolution, it is evolution in the behavioural engine driving the emergence of language.

3(8).15.5.2.2 So – why does language matter to an organism? If it is an attentional mechanism, and the organism is evolving for ever greater complexity of multi-tasking capability, and is evolving in a social context, (a reasonable drive for species differentiation), then that organism's attentional "load", so to say, can be partly managed by language behaviour even as that behaviour contributes to the overall load. The added benefit is that another member of the same species will be able to observe the attentional behaviour, and even learn about the attentional behaviour via its communicative potential (distributed cognition, as per GCP7). Fully fledged dialogue, as we humans know it, is about joint attentional work in the semiosphere.

3(8).15.5.3 This scenario for the emergence of language behaviour carries through to the details of the structure of the language. For example, early vocalizations, if syllabic (as per MacNeilage, 2008) may prove adequate in some phase of the evolution of multi-tasking capabilities – but increasingly inadequate as these capabilities evolve. Therefore, the evolution of the multi-tasking capability must be accompanied by ever finer and more complex muscular control and monitoring, or the SI would not be "fully" served.[4] Part of the SI "demands", so to say, on the evolving organism will of course be ever more complex attentional capabilities. This co-adaptation, which must be part of the potential of any organism for evolution, is necessary – but variable in its delivery one must suppose. The resultant speciation is not always going to be successful.

3(8).15.5.4 In respect of the emergence of human language we could envisage a phase in evolution where unanalysed vocalizations matched the Sequential Imperative requirements, so that, for example, the current complexities of syllabic structure would not be evidenced. There would be no need to control VOT, for example, and possibly no ability either, because that degree of specification of vocalization, with all its attentional implications, and the monitoring required for successful serving of the SI, would not be part of the overall capability of the organism. But evolution is driven by speciation through SI differentiation, with the necessary corollaries in terms of articulatory development, sensory refinement, attentional development, and eventually multi-tasking capability (the latter is not widely evidenced in non-humans, but can be found). So language does not evolve so much as ride the wave of evolution.

3(8).15.5.5 The evolutionary story just told does exhibit something like a discontinuity, but it is not radical. It is the discontinuity we discussed earlier in respect of attention. The account is that evolution in the animal kingdom is driven by speciation through differentiation in the expression of the Sequential Imperative. A new species does things differently – the form of the organism is different, the sensors, the articulators.... everything indeed that is

4 The issue here is that maybe there could have arisen a multi-tasking capability not matched by vocalization potential; brain driven selection pressure for greater vocalization subtlety. The alternative is also possible, of course, that oral/lingual control subtleties existed (for reasons to do with feeding) before the SI drive to exploit them in speech.

involved in serving or expressing the SI. This produces the range of animals we encounter, with a corresponding range of behaviour. But, and here we see what amounts to a discontinuity, managing the SI – the attentional work that an organism must engage in – is itself behaviour, and thus subject to "selection pressure" in evolutionary terms. Part of the SI itself in such circumstances becomes susceptible to the SI driven evolutionary account – attentional behaviour changes through evolution. It is always true that the behavioural engine evolves as the SI drives change – the brain changes, and thus attentional behaviour changes. But at a certain scale of complexity we find that the nature of the behavioural change itself begins to change – attentional complexity is being selected for in evolution. Fineness of articulatory control of the vocal apparatus matters only for speech, and it has to be accompanied by appropriate sensory capabilities, but the constellation of abilities may have been driven by the need to support attentional complexity, and the consequent species differentiation in the SI.

3(8).15.5.6 One interesting side-effect of the system management approach to the SI, and to the suggested conception of communication as attention management, is that the asymmetries in structural properties/ tendencies/ patterns in the structures of languages around the globe may be explainable in terms derived not from "natural semantics" but rather from attention and the drive to use language to get another human to attend to the same thing in the same way.

3(8).15.6 We can now return to the accounts offered by MacNeilage (2008) and Carstairs McCarthy (1999) and review them in the light of the foregoing. Both authors are concerned to point to the syllable as key to understanding speech and language evolution, as they see it. There is no doubting the appeal of the syllable as a linguistic concept, and its possibly crucial rôle in the development of written language. However, in my view these two authors are unconvincing in their arguments that the rôle of the syllable has been to shape language and speech in a way that no other behaviour can, and that, therefore, language evolved from a syllabic core. Where MacNeilage is more plausible is in highlighting the value of the syllable as a basic entity. Recall that cyclicity is a way of providing sequences (1(1).1.4). Mandibular cycles are naturally available (chewing) and when coupled with vocalization and modified breathing cycles can yield syllabic sequences. Interestingly, all that is required for language delivery is fine control of the apparatus – the cyclicity is provided for other reasons (see Tattersall, 2012, on exaptation).

3(8).15.6.1 Carstairs McCarthy (1999) is concerned to show that linguistic structures in syntax derive from the structural detail in syllables. At one point (op. cit.;pp:147–148) he suggests that:

> ... since the syllable appeared as unit of phonetic and phonological organization as soon as the lowered larynx and other vocal-tract changes made a more modern style of vocalization possible, it is reasonable to conclude that the neural organization underlying syllable structure was co-opted to provide a syntax for strings of 'words' when the need became pressing. It was natural, therefore, that syntactic structure should possess features reminiscent of syllable structure. This resemblance was neither accidental nor analogical but rather homological in the evolutionary sense: that is, it came about because sentence structure had originally had the same biological basis in neural organization as syllable structure had.

This claim is unconvincing not just because there are so many languages for which structures derived from the surface ordering of the components of syllables can make no sense (languages with free word order, for example, or the North Semitic languages discussed earlier). Additionally, the account slides over the whole issue of morphology and its complexities – 'words' in the above passage hints at this, and elsewhere Carstairs McCarthy (op. cit.) discusses inflectional morphology. The difficulty he ignores is that morphology is not entrained in any sense by syllabicity. Lastly, the claim does not convince for the additional reason that the sort of co-option envisaged is not feasible. There is no homunculus allocating resources and co-opting structures and processes to serve other needs, let alone one with the foresight required to provide for the eventual "evolution of language" in the generic sense – or even a proto-language in the first instance. He might shelter behind notions of "co-evolution" of language and neural structures, as appears to be the case later in the book, but then the argumentation must falter because co-evolution in the sense he requires is not coherent (it requires matching but separate evolution of both result and mechanism).

3(8).15.6.1.1 Carstairs McCarthy is not the only person to have proposed some sort of account in which a neurological structure of one sort or another is pressed into service for language. Schoenemann (2009) refers to the brain's "circuitry" in these terms:

> The fact that language evolution is constrained by the preexisting abilities of individuals in the population means that any changes in brain

> circuitry relevant to language in a given generation would likely consist of slight modifications of circuits that already exist, rather than major changes in the ways language is processed by the brain. Because this would be true for every generation, language evolution in the long run would necessarily be continually biased toward the modification of preexisting mechanisms, rather than the accumulation of wholly new components (Schoenemann, 2005). As a consequence, we should expect language circuits in modern humans to show extensive homologies with preexisting systems in closely related animals. Even if language evolved to use circuits not originally linguistic in function, these hijacked circuits would likely also represent modifications of nonhuman-specific circuitry. Thus, studying brain and behavior in nonhuman primates is actually central to understanding human language evolution.

Again we have a proposal for change, or development (but note that this is gradual, not abrupt) that looks a bit odd – who or what is doing the "hijacking" of circuits and how on earth would this work? The evolution of a more complex behaviour (species) will involve the hijacking of pre-existing brain circuits as evolution proceeds – so the brain of non-human primates who use stones to open nuts will have evolved through hijacked circuits for some other behaviour in other species, and so on. The story runs back through time, right down to the smallest brain – but without an account of how hijacking works we are no further forward.

3(8).15.6.2 MacNeilage (2008) is concerned with the evolution of speech *per se*, and considers the notion of the syllable to be the basic linguistic entity from which the whole gamut of languages arises. Clearly there is no archaeological record of syllables from hundreds of thousands of years ago. The point is rather that the mandible movement in feeding, which is necessarily oscillatory, underpins the fundamental speech gesture: the syllable. He invokes Haeckel's recapitulation notion in respect of the emergence of speech in children. What we witness in a child's achievement of language – starting with simple syllabic utterances – is a recapitulation of how language is supposed to have evolved in the species.[5]

5 This style of argument also surfaces in a recent paper by Imai and Kita (2014). Interestingly, one could consider a Haeckelian recapitulation argument for modeling the "language as attentional mechanism" approach described earlier. Children's acquisition of language would then be a sort of recapitulation of the evolution of attentional control via language. But in

3(8).15.6.2.1 One troubling aspect of MacNeilage's (2008) account is that he assumes too much too quickly. His notion of frame and content as the basic conception of the syllable is applied too readily to what can only be called pre-speech, despite his expressed caution. He notes (op. cit. p:109), for example, in respect of babbling:

> ... it consists basically of a rhythmic alternation between closed and open states of the mouth, powered by the mandible. Its period of occurrence is roughly from 7 to 12 months of age. The typical babbling episode begins with a consonant-like sound and ends with a vowel-like sound. I use the cautionary terms 'consonant-*like*' and 'vowel-*like*' (which I will immediately abandon for convenience) because the terms 'consonant' and 'vowel' imply independence of the two types of unit in the control process and, as I noted earlier, babbling infants lack such control.
>
> An important difference between babbling and the comparable subsong stage that precedes birdsong is that these consonants and vowels are typically 'good' sounds, totally acceptable as language-like... Only a small set of sounds is favoured – voiced stops, nasal, and glides, at labial and coronal places of articulation, are the main consonants... while the vowels... tend to be either mid or low, front or central vowels... The basic close-open alternation and the favoured sounds change little in the first-word stage, lasting from about 12 to 18 months, during which infants typically produce only about fifty single words, one at a time.

In the second paragraph he abandons the caution expressed in the first and implies indeed that babbling involves the production of recognizable consonants and vowels. There are several difficulties here.

3(8).15.6.2.2 The first difficulty is the assumption that only part of the totality of acoustic evidence for the mandibular movement is important – namely the opening and the "vowel-like" sound – but not any acoustic evidence of the necessary closing. Even here, however, it remains just an assumption that the opening phase is even "consonant-like", let alone an acceptable language-like sound. What is important is the child's intention – and this may be no more than a series of cries ("vowel-like" sounds) without any thought to the details of the otherwise merely necessary mandibular cyclicity used to produce them. In short – it is not clear that babbling involves frames and content at all.

---

fact I think Haeckel's conception brings no insights to further our understanding – it is an appealing but misguided error of judgement.

3(8).15.6.2.3 The second difficulty is that the frame and content model of speech appears to be undermined if the notion of frame and its content is not fixed. To consider that the frame/ content structure itself evolves – starting with, perhaps, just a vowel-like grunt and no control of the release, or supposition that it matters, undermines the model. This in turn undermines the application of the model to the infant's babbling, because the implication is that the infant can learn what to do without utilising a frame/ content structure (which thus remains just an expository convenience). Yet it shouldn't matter, actually, because the evidence is that the release phase of the babbling style grunts, or better, cries, is not controlled. That this portion of the mandibular trajectory can be controlled in respect of what comes to be considered the articulatory configuration (tongue position, use of lips, management of release abruptness, nasality) is beyond the infant's initial comprehension or capability. So it would be better not to invoke a model of the achievement of speech which requires the totality of the syllabic frame and its application to everything the infant does. Likewise, the frame does not need to be held out as that to which the infant aspires.

Infants, in other words, do not babble in syllables, and simply learn to produce noises with ever increasing refinement in the control of the production of those sounds, eventually producing patterns we recognize as syllabic. This trajectory implies that the frame/ content notion has little value. However, MacNeilage is surely right to single out the syllable as the basic unit in language (cf. 2(5).10.3.4.2), even if his notion of "language evolution" and his invocation of Haeckel are steps too far. The syllable – a topic which surfaces repeatedly in this book – is perhaps best seen as the basic unit of linguistic activity because it is the smallest arbitrarily specified unit of behaviour which exhibits duality (GCP2;(ii)) and thus provides for the semiotic exploitation, licensed by GCP1;Cor2, which is required in language behaviour. The issue of its smallness being evolutionary prior, or somehow primitive, is not relevant.

3(8).15.6.2.4 MacNeilage (op. cit.) does cover one rather intriguing aspect of the discussion of evolution not commonly addressed. What is the rôle of hemispheric specialization – seen in most if not all vertebrates – in language behaviour in humans? Language capability in humans is clearly not a consequence of brain lateralization. The impression left with this reader is that what he is actually driving at, in my terms, is some sort of hemispheric differentiation between sequencing and de-sequencing. This remains to be explored, but would if even partly substantiated help support an account of the functionality of the brain in terms of serving the SI. Non-automatic attention and control of sequencing and de-sequencing must be presumed to have evolved very early – the Sequential Imperative is, after all, unavoidable. Lieberman (2006: 181ff) discusses problems with what he calls the "sequencing engine" – the basal ganglia – as evidenced in Broca's syndrome. And his analysis of evidence from speech disruption shows that it is the production of stop consonants which is disrupted, specifically the sequencing required to get the appropriate VOT. Other aspects of speech production (vowel length, for example) do not show comparable disruption. Importantly, the basal ganglia (as a structure in the brain) is early in evolutionary terms. Lieberman does not discuss de-sequencing.

3(8).16 We can begin to broaden the discussion now with a look at signed language and gesture. There have been various suggestions that signed language preceded spoken language in human evolution – that is, that speech evolved from signing (see Armstrong et al, 1995; Hewes, 1978; Westcott, 1974). MacNeilage (2008) is unconvinced and reviews the arguments at length from the perspective of structural dissimilarities, concluding that amodal linguistic control of production is not plausible (for a different but also negative take on "gesture first", see McNeill, 2012). But actually it might be argued that the "language evolved from gesture/ tool-usage/ signing/ younameit…" school of thought is part of what motivates MacNeilage because his own "younameit" is mandibular oscillatory patterns leading to syllables leading to speech leading to fully fledged language. My concern, as should be abundantly clear by now, is that it is wrong to speak of "a behaviour" like gesture, and it is wrong to think that such "a behaviour" evolves into "another behaviour" we call language.

3(8).16.1 MacNeilage (op. cit.) fails to note that in fact there are some arguments for structural similarities between speech and signing in what could be called amodal notions of syllabicity. Arguments about syllables in signed language production have revolved around the identification of signed equivalents for consonants and vowels. However, it has been suggested that this is a distraction and that the notion of sonority makes more sense as the basis for looking for syllables in signing. I proposed (Edmondson, 1986b) that delivery of specificational completeness of the manual component of a sign amounted to the equivalent of the sonority peak. The trough between syllables is located at the moment when the specificational components – hand orientation, hand shape, hand position, movement – cease to have linguistic values as one sign collapses and the hands move meaninglessly toward the next set of specifications. Note that in the transitions the hands do not cease to have any shape or orientation or location or movement – that would be impossible. What happens is that the realisation of specifications or values supplied linguistically collapses to zero (but there may be some anticipatory perseveration of, for example, handshape or orientation; cf. the discussion of lip-rounding in 2(5).10.2.4.2). The issue of syllabicity in signing remains unresolved, and is currently perhaps of dwindling importance in sign linguistics. Nonetheless, MacNeilage's drive to find no plausible signed equivalent of spoken syllables is perhaps not as readily substantiated as he would wish. Further, there is a significant body of work on amodal linguistic characteristics and structures (see Meier et al, 2002) so the general thrust of MacNeilage's attempt to undermine the "gesture first" idea is actually undermined by the work of others.

3(8).16.2 The relationship between gesture, signing and speech is the subject of academic enquiry, but not extensively so. The work is deeply influenced by McNeill (1992, 2000, 2012) but the academic discussions go back to the 1960s and '70s and the start of serious work on signed language research (Klima and Bellugi, 1979; Stokoe, 1960). In the late 1970s (Bellugi and Studdert-Kennedy, 1980; Friedman, 1977; Siple, 1978) there was some concern with the imagistic quality of signing, the nature and rôle of classifiers in signing, and the mapping of sentences onto real-world sequentiality (cf. "the train left 3 minutes after John got on"; "John got on the train which left 3 minutes later"). This concern was contextualised by the debate about the linguistic nature of signed languages – did they conform to Hockett's design features (1960; cf. Schlesinger, 1978)? Did they share typological characteristics with spoken languages (use of classifiers, possible typological categorization as polysynthetic or incorporating, and etc.)? The underlying motivation was to show that signing is not just elaborate gesturing (although that begged a whole slew of questions anyway). Of course, the imagistic nature of signing need not fight claims about syntactic complexity, linguistic status or whatever. Around 35 years ago I was intrigued by these worries (I was working at the time on "informal sign language" use in some schools in London with an oralist philosophy (see Introduction to Part 2) – and found that in fact the "language" used by the children I studied was somewhat impoverished, as there were no adults using British Sign Language (BSL) to provide linguistic models (Edmondson, 1981)). I explored a simple experimental idea (nothing was published) which showed that in some circumstances it was possible for adult native signers of BSL to comprehend filmed signing shown in reverse motion. NGIS, as I called it, was briefly a subject of some humour at conferences I attended, but it was noted at one such that an attempt to create "nonsense" signing for some psycholinguistic experiments, by means of "backwards signing" (samples played in reverse motion) failed because there was too much comprehensibility in the stimuli (the signed language concerned was not BSL or American SL; probably a northern european SL – I don't recall which). When relaying my observations to a colleague[6] she responded with words to the effect that the subjects had no choice but to produce sequential movements, even though the order didn't matter. I realised that she was right, of course, and then I realised that, over and above her summary of the behaviour of Deaf people signing, the comment applied to all movements, and all behaviour, linguistic or otherwise, in all animals. I formulated GCP1 on the basis of that insight, and the rest followed over the course of 35 years.

3(8).16.3 Signing is imagistic in many ways – not just the generalised notion of image building by waving the hands around "depictively", or the syntactically constrained morphophonemics of actual signing, which can even, on occasion, be "read" in reverse motion. One component of the image building is the use of classifiers, sometimes called "proforms". These are conventionalized hand-shapes used to depict or "stand for" a noun (e.g. person, or car, or plane) which can be moved continuously in 3D space around the signer (violating syllabicity) (e.g. see Mandel, 1977). Although it can be argued that classifier usage can be accounted for syntactically within a framework of discontinuous morphological expression (see Edmondson, 1989) this does not counter notions of imagistic use of classifiers. Image building in signing can be really thoroughly constrained by convention and syntax and morphophonemics. Indeed, it is my view that the use of classifiers in scenario situations – e.g. when talking about a car going on a long trip along a road which winds left and right, up and down hills, and etc. – is entrained gesturing (which otherwise is thought not to exist in signing).

3(8).16.4 Gesturing whilst speaking is the topic that McNeill's work has done much to promote. One conception in this work is relevant here – namely that somehow a gesture manifests rather directly a core cognitive representation (Growth Point) underpinning language and from which the spoken utterance is derived (this is not sequentially presented – it is precisely coordinated and synchronized with the gesture). This could be couched in a sort of Haeckelian framework applied to behaviour; ontogeny recapitulating phylogeny in the sense of gesture evolving into speech – but this is not done in his work (cf. the earlier account by MacNeilage of syllable evolution).

6 Dr Monica Lawlor, Psychology Department, Bedford College, University of London.

The McNeill account of speech and gesture is, to be fair, more complex than just characterised. He writes of the Growth Point (GP), (McNeill, 2012, p:19):

> To answer the question, Why do we gesture?, this chapter says that it is an integral part of thinking in language; that combined with speech it creates a dynamic dimension on which thought and speech come alive. Observing the gesture and the co-expressive speech it synchronizes with, we witness the ongoing imagery-language dialectic of which the GP is the unit.

The work is difficult to align with more conventional accounts of linguistic activity (which do not cover gesture) but makes it clear that it is probably better to reconceive language as coupled with gesture, than to attempt any sort of "add-on" approach to bringing gesture into linguistic discussions. Nonetheless there are problems; the GP account of gesture accompanying speech presents some difficulties for the emergence of signing in Deaf communities where the language-gesture dialectic involves behaviour in a single modality. If gesture is somehow a necessary (but separately controlled) aspect of spoken language, then signing (which seems incontrovertibly related to gesturing) looks problematic. More to the point, of course, and as McNeill would wish, stories about the sequential emergence of different modalities (gesture before speech) become rather difficult to sustain.

3(8).16.5 One criticism of the "gesture before speech" story about the emergence of human language – that it is really odd to invoke the evolution of a modality switch, with all the attendant supporting functionality – is surely as valid in relation to McNeill's account of gesture as it is in the explicitly "gesture first language evolution" account. McNeill's account seems to imply that language evolved twice in the one species. It is more parsimonious to suppose that gesture is part of language behaviour – or more generally communication behaviour – with the "sequencing engine" producing outputs in more than one modality. It is not clear why it should be thought that the involvement of more than one observable modality should imply the operation of more than one underlying activity. Thus, to argue that the GP is the primary entity injected into the "sequencing engine" from which two streams of surface activity finally emerge is unnecessary. Better to conceive of injection into the sequencing engine of the appropriate behavioural precursors (entities) for the desired observable behaviour – behaviour which could include, incidentally, much more than just manual gesturing and speech itself (for example, posture, direction of gaze, facial expression, tone of voice, attempted mimicry and etc.).

3(8).17 Before turning to our last chapter we need to summarize our thoughts on evolution *per se*. The argument is clearly that language itself does not evolve, but rather that the behavioural engine evolves – opportunities for, styles of, and complexities in the processes of sequencing and de-sequencing (which cannot be avoided) change as brains evolve. The argument presented for language assumes that it does not make any sense to think of evolution of any behaviour at all, let alone evolution as some sort of "progress" from "primitive primate" to *Homo chatterbox*.

3(8).17.1 Briefly, the story being told here is that the brain, any brain, is compelled to serve the Sequential Imperative. Small brains may accomplish this in rather inflexible style with automaticity of attention and control of processing. This does not imply a total reliance on innatist accounts – small brained creatures do learn patterns and actions (bees, for example, learn about the disposition of food resources, Gould and Gould (1995)). Brains necessarily co-adapt or co-evolve with their host organisms. If a sensory organ increases in size or complexity then (part of) the brain must co-evolve to match that in order for functionality (the SI) to be served. This is thought to have been the case with *Homo neanderthalensis*, with the possibility that the expansion of brain to serve vision (larger eyes) came at the cost of brain availability for other behaviour (Pearce et al, 2013), given that the cranium cannot simply get ever larger under evolutionary pressure because (a) there will be birthing difficulties and (b) larger brains require more food and more successful food processing in the host animal (as discussed in the literature in respect of human evolution; see e.g. Tattersall, 2012). Likewise, as the potential for multi-tasking increases – say through use of forelimbs for tasks other than locomotion, along with fine control of the hands on such limbs – then (part of) the brain must co-evolve not for size, but for complexity of action in order for the SI to be served (and the potential for multi-tasking to be realised).

3(8).17.2 So what does evolve? The Sequential Imperative is the functionality which must be delivered by all brains. As organisms evolve, the potential for and style of serving the SI changes (speciation). Beyond the merely necessary changes to match the demands of the SI to the body (increased size) there are two other categories of "demand for more brain", already mentioned. One is increased sensory apparatus demands – innervation attached to larger or more "refined" organs must necessarily terminate somewhere in the brain and thus intra-species co-evolution is inevitable. The complexities of avian vision provide an illustration. Some birds have eyes with two fovea, with stereoscopic vision not as overlapping in fields of view as we find for ourselves (it may indeed be absent). This is coupled with relatively large eyes (in relation to body mass mostly larger than in mammals) very close to the visual processing centres in their brains. These factors all "conspire" to enable birds to display behaviour which seems magical. Apparently unbelievable acuity coupled with rapidity of response to stimuli means that birds seem effortlessly able to track distant predators, pursue prey, fly fast with extraordinary manoeuvrability, flock in close proximity to thousands of other birds (starling murmuration provides the most dramatic display of flocking), and so on (see Birkhead, 2012). "Bird-brain" is thus a mistaken insult – the brains of birds are very well adapted to serve their (SI) needs.

3(8).17.3 The other "demand for brain" comes from articulators – for example complex configurations of articulators where independence of activity within the configuration requires independence of control and thus more complex delivery of the SI. This can be seen in hands and limbs, and mouths, and in combinations of these. Birds (again) are good at fixing something, for example with a foot, whilst they operate on the item with their beak. Furthermore, avian tool use can be quite complex (see Taylor et al, 2007). Various primates can use their hands independently of their feet, to wash potatoes, fish for termites, wield stones to smash open nuts, and so forth. A horse has learned, apparently, to manipulate its stable door fastenings with its mouth/ tongue, so as to get out of the stable, after which it can open other stable doors and pretty much go where it pleases (the horse is called Mariska – a YouTube star in early 2013). The ability of an animal to manipulate its environment with ever finer control, to acquire food and/or to manipulate the food in its mouth, would seem to have distinct survival value – and hence be a driver for speciation through demands for "more brain".

3(8).17.4 Clearly this is loose talk – sensor/ articulator evolution cannot demand changes in the brain. However if, say, articulator development takes place, then secondarily changes in the brain may make use of the articulator development through exaptation to produce, say, greater dexterity. Or indeed *vice versa* the brain may develop an increased repertoire of processing, or of sequencing and de-sequencing capabilities, which might surface initially in more comprehensive control of limb articulation but which, as articulators develop, emerges in addition as refined dexterity.

3(8).17.5 We have covered a considerable amount of territory now. There are perhaps two major issues arising from this chapter. The first is that language does not evolve from other behaviour(s). What evolves is the engine for serving the Sequential Imperative. This is necessary because changes in any organism which yield different sensory equipment or articulatory potential have to be aligned with changes in the brain to make those "peripherals" work. Speciation is the consequence of the emergence of such changes/ differences. The second point is that as the delivery of the SI becomes more complex the changes required to make it all work become internal, rather than concerned with sensors and articulators. Necessarily, it seems, the management of the SI looks like more behaviour and that behaviour (following the line from attention getting in alarm calls, through to the organization of social cohesion in social species) is focussed on delivery of attention in complex multi-tasking scenarios. The emergence of language, in short, is the necessary emergence of attentional control of the SI in a modality which is not obstructive or conflicting in respect of other behaviour. This could be studied in some depth in further work, but to provoke the reader a little – language may not be the only behaviour deployed by brains seeking to control the complexities of attention in managing the SI. We will consider this in the next chapter.

Additionally, we hold over to the next chapter discussion of evolution in the (de-) sequencer and the BGPs (see Figure 5.7) and implications for the management of the Sequential Imperative and thereby for the emergence of language. We can also consider another topic – consciousness – which appears to be intractably linked to the brain (irrespective of the sensorimotor interface). So, our next chapter is concerned with more general questions focussing on the possible rôle of language in thought, the nature of consciousness and the interesting question: what is it like to be an ETI?

3(8).17.6 What can we say now about the emergence of language? We need a thorough account of the evolution of the (de-) sequencer and the background processors (and see Chapter 9). We need to combine that understanding with some sort of account like the one offered in Jackendoff (2002; Pt2), but without UG. This is another big project – indeed maybe the biggest project – implied by the account of the Sequential Imperative presented in this book.

CHAPTER 9

Language and Consciousness: What Is It Like to Be an ETI?

3(9).18 In this chapter we set out to achieve a variety of goals, including bringing the arguments together to show that the sketch we set out to provide at the start is complete enough for others to work on. The sketch will also illustrate the idea that we can say something about what it is like to be an Extra-Terrestrial Intelligence. But before we get to that point we need to extend our evolutionary account briefly, to cover non-human primates and non-linguistic management of the Sequential Imperative. Our functional model of the delivery of the SI has some implications for evolution of the brain in terms of functionality – we need to discuss these. We also need to consider the rôle of language in thought, and the nature of consciousness.

3(9).18.1 It would seem that humans have evolved a complexity of multi-tasking unrivalled amongst other animals. This is behavioural capability which arises through general speciation pressures operating on evolution of the brain/ sensor/ articulator complex serving the Sequential Imperative. But new behavioural capabilities arise which are also merely extensions of the basic repertoire required for delivering the SI (for any given sensorimotor-corporeal interface). What is evolving, in short, is modelled both by the number of tiers/ half-planes (or pages) in the (notebook) model discussed earlier, and by the nature or repertoire of background processors. The functional model which has served us well enough thus far highlights the need to recognize that the actual wetware may be quite complex (!) in that different domains therein may evolve separately. If there are wetware equivalents in some way of the (de-) sequencer, and of the BGPs, then these have to be considered as separate domains. Notice here that the functional model of the delivery of the Sequential Imperative tells a complex story about what is evolving; we will assume that the model is saying something important about the brain. Notice further that both the model and the account we develop do not invoke or require any form of modularity in the sense encountered in Cognitive Science (see Fodor, 1983; also 2(6).11.3).

3(9).18.1.1 Increased opportunities for sequencing and de-sequencing, and for managing this process with ever more elaborate attentional behaviour

© KONINKLIJKE BRILL NV, LEIDEN, 2017 | DOI 10.1163/9789004342996_014

(tool-making and using), drive speciation in primates and the emergence of social activity (e.g. grooming) and eventually language. Language is here viewed as just another attentional mechanism and provides for management of the attentional behaviour. Early signs of such management will be vocalizations as attention management behaviour in various species – for example, alarm calls – and then the differentiation of these according to predator, and even according to the predator's behaviour (see the discussion of the shrike (*Lanius excubitor*) in Birkhead, 2012).

In humans, as the attentional work becomes more complex, with ever extending "now-spans", the requirement grows for language to manage attention displaced in time, and space, from the current moment. A grunt and a pointing gesture are fine for the moment. Something more complex is required to "point out" that the animal being watched, for example, is the one wounded yesterday. Thus, language behaviour actually becomes *necessary* for the management of the Sequential Imperative as that management becomes more complex. However, along the way, so to speak, other behaviour may have sufficed for managing aspects of attention in both individuals (tool making/ using) and in groups (grooming). And in addition there will be differences between individuals within the species (no homogeneity is required or assumed here). SI management is therefore a more broadly varied activity than is suggested by the account in Chapter 7.

So, set alongside the complexity of management which requires the use of language in managing attention and thus the SI (in humans), the breadth of management activity should not be understood as implying the inevitable or ubiquitous use of language in SI management. Other forms of behaviour can serve this rôle (tool making, image manipulation, action rehearsal) as we will see later in discussion of consciousness.

3(9).18.1.2 The evolution of multi-tasking must be limited (and see 3(9).18.3 below). The problem with envisaging unstoppable emergence of ever more complex multi-tasking (used here also as a proxy for discussion of intelligence) is that it all has to be managed (see Chapter 7). There will be individual variation in multi-tasking capability, possibly based on experience but perhaps also genetic endowment (cf. the implications of the discussion of cultural transmission in primates of the use of stones to break open nuts, 1(3).7.4.4). But it seems inevitable that limits on the management capability will be supplied by the requirement to ensure the SI is served in an organized way. It is as if the number of half-planes in the model cannot be increased without limit. Evolution of the brain (here on Earth), it seems, "tops out" in *Homo sapiens sapiens*.

3(9).18.1.3 Contextualization is also important in multi-tasking. We saw earlier (1(3).7.4.5) that contextualization matters in discourse management – a form of multi-tasking. We also saw that both Projected Cognition and Distributed Cognition are involved in discourse management. An evolutionary perspective on multi-tasking must include the idea that Projected Cognition – essentially private, but nonetheless involved in managing the delivery of the SI – appears to evolve into Distributed Cognition. As already noted, behaviour does not evolve. What we have here is the evolution of the multi-tasking capability which comes eventually to facilitate DC and thereby within an individual the move from PC to DC in some circumstances.

3(9).18.2 In terms of "language evolution" in the *cultural* sense – it seems entirely possible that this has happened several times as groups of early humans (*Homo sapiens*) spread out around the planet and coped with the demands of their environments (see Nichols, 1998, for a discussion of early language families, but note Trask, 1999). Whether this means that "proto-language" was a one-off development, or also occurred several times, remains to be worked out (cf. Chapter 8); the challenging question is likely to be what is really meant by proto-language.

3(9).18.3 The emergence of *Homo sapiens sapiens* with large (but unfinished) brains is limited by both birth canal restrictions and metabolic requirements. The plasticity[1] in the brain's structure provides well for adaptation to the environment, learning, cultural transmission and etc., and would seem to be a fortuitous response to the problems of delivering more brain for the adult, but in a limited delivery system, with limited metabolic resources. But there is one additional factor of interest – idiosyncratic development to the adult form means limited homogeneity in the population. This may provide a route to cultural divergence, but also a limiting factor in the development of ever greater brain size/ complexity. Indeed, as the complexity of provision via plasticity increases, the drive for Sequential Imperative driven speciation grinds to a halt once language can be taken for granted. In a sense, cultural "speciation" mediated in part via language differentiation inevitably replaces genetic speciation.

1 The Critical Period hypothesis for achieving language (see Chapter 4 of Lenneberg, 1967, and the ensuing debate in the Cognitive Science literature) has it that achieving one's first language is dependent on neuroplasticity – the fact that the brains of humans continue to develop for some years after birth (perhaps as far as puberty). In other primates the period is shorter, and of course they do not achieve language, but there could be critical periods for aspects of behaviour more widely than just language.

Conventional evolution then appears to have stopped in *Homo sapiens sapiens*. Maximal multi-tasking capability (and its management) does not have to be expressed in only one way – individual variation develops from birth (at least) and in cultural contexts. Selection pressure in evolutionary terms is dissipated.

3(9).18.4 There is one further aspect of the evolution and language story that needs some reflection. In a well-known paper (Nagel, 1974) Thomas Nagel asked the question: "What is it like to be a bat?". The difficulty one has answering that question points, supposedly, to problems with reductionist explanations of consciousness, of mind, and perhaps also of evolution more generally. Nagel's recent thinking on this topic (Nagel, 2011) makes this clear – there really is, he seems to believe, an insurmountable problem for reductionists who want to provide within a single scheme some explanation of the mind-body problem, consciousness and evolution. He considers that his informed incredulity concerning evolution provides a plausible premiss (to accompany his understanding of what the mind is) to work up an argument to the effect that we can never know what the mind is, or understand consciousness in humans and other animals, or how we got to have these, through reductionist accounts of the world.[2] He is aiming for some sort of post-materialist account which accords mind a natural status in the scheme of basic explanation which can be used to account for thinking organisms. This is not to say that mind is like earlier proposals for the aether, rather that its extra-materiality needs to be invoked when accounting for beings with brains, alongside materialist accounts of the material aspects of the organisms.

3(9).18.4.1 The relevance of Nagel's work here is threefold. Firstly, his concern about the believability of evolution is another prompt to reflect on alternatives to "language evolution", although in this case the worry is that a very different sort of explanation for language will have to be found. Secondly, Nagel says almost nothing about the emergence of language or its rôle in the mind-body debate, the functioning of mind and cognition, etc. His approach might be associated with special case arguments separating humans (with language) from other animals (without), with Universal Grammar perhaps being evidence for the post-materialist view of minds in organisms. But he doesn't venture any of this (it seems likely someone will attempt this). Thirdly, he is as concerned in the recent work as he was much earlier with the notion that his accounts of problems with consciousness, subjectivity, etc., have universal applicability.

2 Note that the attempt herein to elaborate a functional specification of the brain is essentially reductionist.

For example, in the 1974 paper he points out that the bat would have a problem understanding what it is like to be a human, and that a Martian would also have such problems (it appears he invokes a Martian on the grounds that the bat could be excused through its limited intellect). Martians get several mentions in the 1974 paper. In respect of citing Aliens it seems that his thinking leads him inevitably to want to assume that the principles he is uncovering have universal validity.

3(9).18.4.2 Another characteristic of Nagel's work is its lack of definition of such concepts as mind, consciousness, cognition, etc. This is hardly surprising, all such work is exactly an attempt to pin down these subjectively self-evident notions and experiences. In the next section of this chapter we will approach this definitional challenge in a different way by attempting to find aspects of the brain's delivery of the Sequential Imperative (and our model of the SI delivery) which match the subjectivities. The eventual success or otherwise of this is perhaps less important than the approach itself. We now continue to examine the account provided by Nagel, but this time focussing on consciousness.

3(9).19 In 1974 Nagel began his paper "What is it like to be a bat?" (Nagel, 1974) with the sentence: "Consciousness is what makes the mind-body problem really intractable". In recent work (Nagel, 2011) we find, in the section on consciousness, the following opening sentence: "Consciousness is the most conspicuous obstacle to a comprehensive naturalism that relies only on the resources of physical science". In what follows I will address the problem of consciousness and propose a rather radical solution to the mind-body problem.[3] But before doing so it is necessary to explore the subjectivities entailed in answering the question "What is it like to be a bat?", and we will do that by asking "What is it like to be a human?".

3(9).19.1 To tackle this question I will divide the "internal world" into three components and treat them separately for the purposes of clarifying some issues. Two of these three components have already been alluded to in the discussion earlier of the model captured in the diagram in Figure 5.7. We have (a) the central spine of sequenced entities of varying durations and including the proximal internal stimuli from the senses, and internal precursors or specifications for the details of articulation – along with the half-plane processing

3 Nagel's views on the difficulties of getting to grips with consciousness can be contrasted with a radically different, neurobiological, perspective in *The Ancient Origins of Consciousness: How the Brain Created Experience*. Feinberg and Mallatt (2016).

for attention which also expresses the differing "now-spans". And (b) the background processes (BGP) which operate "off-line", so to speak, both supplying data to the spine via the half-planes, and "reading in" data from them. What is missing in this model is the sensorium, although it can be envisaged as a component of the BGP in some sense – this will be elaborated below.

3(9).19.2 As a human I might be supposed to feel justified in claiming something like a privileged position in answering the question – I know what it is like to be a human because I am a human. This then isolates my knowledge of my own species from attempts to "know what it is like" to be a member of any other species – and the failure in this wider endeavour (a necessary failure it would seem) undermines knowability in the general sense we would expect to be feasible (in some measure) in a reductionist account. But this chain of reasoning actually doesn't work – partly because my knowledge of what it is like to be a human is in some ways not much better than knowing about other species, as we will see. It also doesn't work because the fact of the Sequential Imperative supports a sense of knowability (knowing what it is like to be an Alien, for example) which goes some way to encouraging belief in the possibility of an eventually complete reductionist account.

3(9).19.2.1 Humans believe they "know what it is like" to be human because they share understanding of some skills, or some experiences of the world, or whatever. I can drive a car as can many other members of the species so surely there is at least a part of that common experience which is shared, and thus knowability is established? The problem with this line of reasoning is that it takes too much for granted. The commonality of behavioural skill – as demonstrated – does not indicate commonality of "internal world" experience. As any child knows – the greenness of grass as they see it cannot be known to be the same sensation as the greenness of grass as their pal sees it. The sensorium is intensely personal (more on this later). In fact, as was pointed out much earlier (1(2).5.1.1) the only aspect of experience we can guarantee to be the same in "shared experiences" is sequentiality, which penetrates the corporeal boundary. The "A before B before C… – ness" of the world is faithfully internalised in that sequence for all of us. What happens thereafter, in the de-sequencing, is individual. Our internal worlds really are just our own even if the fact of having an internal world is understood to be commonly experienced.

3(9).19.2.2 The notion of distributed cognition (GCP7), discussed earlier (1(3).7.4) and to be revisited shortly, might also be thought to indicate some sort of knowability of another's mental world. When we experience those occasions surely there is a fundamental knowing of minds – mutual "mind-reading" so to speak in the joint accomplishment of the task at hand. However, what is shared in such experiences is knowledge of the task and its requirements as behaviour, not as mental states. The fact that I might hand an implement to a friend, because it is closer to me than to him, and I can see that he needs it to tighten up a part on the motor on the lawnmower, or whatever, indicates that I can "read his mind" doesn't it? Not really.[4] What we know of others through demonstrable skill is really rather limited. The fact that my emotional state might include irritation that he started the job without ensuring the tools were to hand is not necessarily recovered through observation. Likewise I cannot recover the thoughts and emotions of my friend who might be very distracted by the thought he had when starting the work that the parts were so worn he might not be able to effect the repair again, and needs to consider buying some replacement parts. Understanding task structure is one thing – mind-reading is something else.

3(9).19.3 The important thing about shared understandings of tasks and behaviour is that the externalities provide for common ground and thus anchor the sense of distributed cognition. The facts of sequencing and de-sequencing are understood – as if tacitly we all knew about the Sequential Imperative without ever putting the assumption into words. However, what we call the mental world is actually composed of other types of "constituent", in addition to task structures and procedures. One of these is sensation – the colours, sounds, shapes, feels… of our experience – often referred to as qualia, or as existing in the sensorium. These, it seems, cannot be shared or co-experienced. But having them is surely enough for me to know that others have them and thus I can know what it is like to be human in that respect at least? Possibly not.

3(9).19.3.1 What is troubling me here is the observation that my coloured grapheme form of synaesthesia is unique to me (synaesthesia appears always to be unique to the individuals who "have it" – whether coloured graphemes or shaped flavours or coloured music or whatever). The yellowness of the printed numeral 8, for example, is something I've never been without – likewise the greenness of 3, the orangeness of S, and etc. These colours are ever-present

4 "Theory of mind" work with children shows that they eventually come to understand what another person must know about a situation (location of hidden objects, for example). This is not mind-reading, but it is reasoning about what it is like to be a different person. For an early overview of theory of mind work see *Theories of theories of mind*. Carruthers and Smith (1996).

and simultaneously experienced with the black and white of the conventionally printed (or hand-written) characters.[5] Discussions with other synaesthetes reveal some shared experiences (see also Edmondson, 2002), especially the experience of trying to explain the sensations of synaesthesia to those who don't "have it". But the incidence is about 5% of the population or less – which means that there is a sense in which I can never know what it is like to be 95% of the population of humans. Synaesthesia is not learned – so a few people with one or another form of colour blindness might still have coloured graphemes, say, but a much impoverished sense of the colours of objects and the environment. All of which is to say – the experience of qualia, in the general sense, we possibly know we share with all animals which have senses. Anything more specific would seem to be irrevocably idiosyncratic and not shared.

3(9).19.3.2 In the diagram Figure 5.7 there is an undifferentiated component referred to as the BGP – back-ground processes. These are described as being "off-line" in the sense that although activity is going on – entities are being accepted as input (de-sequenced) to be stored (memories), or recovered for externalising via the sequencer – the temporal flow is not aligned to the real world. To be sure, if I learn a fact I did not previously know there is a "before" and "after" sense of time in the BGP but that sense of change or development or learning isn't obviously linked to the flow of time in the external world. Indeed – there may be many such changes in BGP going on at the same time and at different rates (involving entities at different scales) – and this is why I use the term "off-line". This characteristic is most starkly brought home to one when violated – because just occasionally there is indeed a link to externalities. So-called flash-bulb memories are strangely complete, including emotional state; they are focussed often on quite brief moments of intense significance (for my generation the "what were you doing when you heard JFK had been shot?" moment).

3(9).19.4 What I want to suggest now is that the BGP in the model is more complex. There would seem to be two completely different parts of the off-line domain. One is the sensorium – the part of the system which lights up, as it were, when sensations are evoked – either by external stimuli or internally for one or another reason (synaesthesia, voices in the head, whatever). The sensorium would have to be invented if we did not already want to argue for its existence because the monitoring aspect of sequencing, via de-sequencing, needs (proximal) evidence, and sensations provide such evidence.

5 Coloured text is weirdly double-coloured for me and other coloured grapheme synaesthetes.

3(9).19.4.1 The sensorium does not engage in processing in the sense we think of for the BGP – off-line creation or manipulation of cognitive entities and their componentialities. However, it is also not the case that the sensorium is some sort of passive repository for remembered sensations. There is an important sense in which the experienced qualia are not very directly aligned to the workings of the sensors. In vision, for example, the sensation of colour is not obviously derived from relative strengths of stimulation of three different colour receptors in the eyes; and although a static visual scene may be seen via a succession of saccades (see 1(1).4.1, 1(2).6.4, 3(7).13.4.1) the proximal internal representation is of the static scene. The sense of the sound world of bird song in a wood, or orchestral instrumentation in a concert hall, is not self-evidently a complex acoustic pressure waveform striking the ear-drum. We hear blackbirds and robins, horns and violins. We learn to identify these sensations, but they are already very indirect even when we don't know what we are seeing or hearing. Sensations have a proximal feel to them, a directness (which we know to be illusory) but they are also inscrutable in a way in which the mind is inscrutable, as we discuss below.

Importantly, sensory stimulation yields sensations which do not have components. This means that the sensorium – a part of the BGP – is about learning and recognition of "stimuli" but we know that these stimuli are already processed to be significantly distal interior representations rather than proximal sensor outputs. The extent to which they appear de-sequenced "going in through the skin", so to speak, reflects links with other de-sequencing happening at the same time. But sensory de-sequencing does not match other de-sequencing because we do not (cannot) sequence sensations for output; the sensorium cannot project into the (de-) sequencer (and thus it plays a limited rôle in the functional account of the brain). Indeed – that may be the distinguishing feature of qualia. Pain may be a sensation without input de-sequencing because (a) it is readily generated internally (kinaesthetic feedback is too, but is linked to muscular activity); (b) it is informative about states and thus provides input to homeostatic control.

3(9).19.4.2 The Sequential Imperative could not function without the sensorium; mimicry would not be possible; rehearsal and refinement of action could not take place. It is envisaged, in this account, that in the spiral notebook model we have not quite abandoned the sensorium is linked into the spine via half-planes (tiers, cf. Figure 5.7) and that it contributes characteristics – we call them sensations – to the entities in the half-planes (it doesn't contribute entities, or manipulate them). The contribution of sensation is triggered from the spinal entities (and this is how the sensory "apparatus" links in) permitting stimuli from the external world as well as internally generated stimuli to trigger sensations in the same way. Evaluation of the sensations can be used for monitoring of sequencing – kinaesthetic feedback, pitch of the voice when speaking, etc. Note a presupposition of sensational isolation except in those cases where synaesthetes combine sensational responses. That flavours might have shape, or shapes might have colour, suggests some cross-talk in the sensorium; and yes, this does mean that the definition of sensation includes shape (think of the Kanizsa triangle illusion). Indeed, I would argue that duration is sensed as well. We will elaborate our thinking about the sensorium and its subjectiveness a little later – after we have considered another way in which the model's BGP conflates complexities.

3(9).19.5 The other part of the off-line domain, the BGP, is what we might call "mind" – a poorly defined term. I am going to define it in relation to the serving of the Sequential Imperative. The functional specification of the brain is to serve the Sequential Imperative – to couple the details of temporally structured perception and action into the atemporal recesses of the brain, so to speak, where goals, beliefs, etc., endure and have structure, but do not have temporal structure aligned to the flow of time in which the being with the brain has to dwell (we will consider "thinking" in a moment). I want now to suggest that the "mind-body" problem is actually just a statement of the fact that the "mind" is not entrained by real world time, but the "body" is. The supposed irreconcilability of the two components of being (and traditionally of being a human) is actually just a statement about the dimensional difference between the two domains. How can this be made to work?

3(9).19.5.1 In terms of the complex notebook style model we seem destined never to jettison the mind is part of what is labelled there the BGP – background processes. In a much earlier outing for Pantome the BGP was labelled the BGC – Big Grey Cloud (but less frivolously, background context). BGC was apparently coined by IBM staff who referred to their remote senior management as the Big Grey Cloud, but did so using a TLA (three letter acronym) for which IBM is apparently well known. The remoteness and "out of touch" characteristic of some senior managements is actually not a bad analogy for the mind in the sense that the processing going on there is not really available to the staff who simply have to clock in every day and get on with the job of implementing whatever directive lands on their desks, and feeding back reports to senior managers about the success or otherwise of whatever they have been doing. These reports will be considered and acted on by those managers – *in their own time*. They exist in their own "little world", so to speak, uncoupled from the daily reality of delivering products and services. Interactions between the managers, and the managerial processes themselves, are not available for scrutiny – that is not their purpose. Managers deliver and interpret and this is what the rest of the staff experience.

OK – so that pushes the analogy a bit far – and I'm not going to develop a society of mind notion modelled on homunculi demonstrating corporate management ability, so let us leave it to one side. My reason for mentioning it is that it captures the idea of communicating background processes which in addition to their main rôle also traffic data into and from the (de-) sequencing activity. The background processors involved – mind – project data into the (de-) sequencer, and at the same time these processors read data from the (de-) sequencer. These data are the consequence of, and also input to, the background processing and inter-processor interaction.

3(9).19.6 Attention would seem to be key to the sense of sensation – and I'd like to suggest here that the multifarious sensations linked to attention with the range of "now-spans" operating in normal waking behaviour, including the sensations which must accompany attention itself, amount to the experience of consciousness. Consciousness is, in this model, the experience of the operation of the (de-) sequencer – it is the experience of multiple attention "now-spans"; in other words it is the experience of (part of – see below) the mind-body solution! I think Nagel's view of consciousness – that it makes "the mind-body problem really intractable" – is flawed. The inscrutably subjective element of one's mental world is the sensorium, not the so-called mind-body problem. It is also true of course, that experiences and events – remembered and learned – are idiosyncratic/ personal/ subjective, but here there is always the possibility that interpersonal differences can be explored through one or another form of behaviour (whether discussion, or sports training, or mimicry, or whatever).

3(9).19.6.1 Further, the "what it is like to be an X" train of thought – *knowing through consciousness what it is like to be a human* (*because I am*), *but simultaneously* (*because my consciousness is not like a bat's*) *not knowing what it is like to be a bat, and indeed never being able to know* (*so mind and consciousness are not reducible to physical accounts*) – this train of thought is flawed. The flaw is actually multifaceted. Firstly, the sense of knowing amongst con-specifics arguably arises through the functioning of the (de-) sequencer – it is the sense that the potential for and exploitation of Distributed Cognition are sort of obvious. But we see this operating in other species, and they even do it with us / for us (the guide dogs for the blind example mentioned earlier (1(3).7.4.1.4)). And indeed one can witness behaviour in other animals which seems to suggest that we humans are more "in tune" with other species than we might think (a ewe will stamp a foot if irritated; a cat might shake a front paw if its mouth touches food which is too hot).

3(9).19.6.2 Secondly, if it is to be argued that consciousness is species specific, and that is the basis for the unknowability of what it is like to be a bat (or Martian), then there must be a physical basis for the difference, and this must (eventually) be recoverable in reductive accounts. One cannot maintain simultaneously that mind is somehow "other stuff" and then simultaneously say its existence depends on DNA – which is perhaps why Nagel tries to argue that evolution is not sufficient to account for the emergence of mind and consciousness. But the fact of physical difference in hosts for minds and consciousness gets in the way – the claim about unknowability actually implies a vital rôle for physical difference and that must entail, eventually, a successful reductive account and some sort of knowability.

3(9).19.6.3 Alongside the sensorium – believed to be experienced by/ in all beings, but unknowable (currently; plausibly some astonishing refinements in neuro-surgery and instrumentation might alter that assumption) – we have the mind as also unknowable, within the human species and indeed within a single individual human. The fact of off-line processing and storage need not be contentious. We become aware of the mind as soon as it projects into the (de-) sequencer (the operation of which we experience as consciousness). This generates behaviour, some of which reaches the external world but some of which does not. The (de-) sequencer provides a way for parts of the mind to couple to other parts of the mind (in real time, and with the contribution of attention (also projected into the (de-) sequencer)). Cognition, on this account, is thus the manifestation of the mind-body solution – the operation of the (de-) sequencer. Earlier it was suggested that the experience of the (de-) sequencer is consciousness – but this was qualified to be the experience of the multifaceted attentional "now-spans" in the operation of the (de-) sequencer. The careful wording is to allow for the operation of the (de-) sequencer – cognition – of which one is not consciously aware (no "now-spans"). The railway carriage balancing trick is part of this, as are aspects of processing such as priming – on the one hand – or on the other hand quotidian behaviour such as pulling out one's car keys whilst talking to a friend, "subconsciously" signalling a desire to move on, and the type of behaviour noted by Freud in the Psychopathology of Everyday Life (Freud, 1901, 1914).

It is worth noting here that indeed Freud viewed the mind as unknowable or inscrutable. Solms and Turnbull (2011) put it this way:

> We can only know the mind via our phenomenal consciousness, which provides an indirect and incomplete *representation* of the mental apparatus and its workings. The actual ontological nature of the mind is something epistemologically unknowable: it necessarily lies behind (and

> generates) conscious perception. We can, of course, *infer* its nature from our conscious observations and thereby 'push back' the bounds of consciousness, which is what the psychoanalytic method seeks to do. Ultimately, however, we can never *directly* know the mind itself. We must therefore have recourse to abstractions derived from inferences and built into figurative models: metapsychology.

Within the (metapsychological) scheme I have outlined in this section of this chapter we see that the mind projects entities into the (de-) sequencer. The mind also derives entities from the (de-) sequencer (in addition to its "off-line" creation and manipulation of entities). In this scheme the mind does not generate consciousness. As noted above, the operation of the (de-) sequencer is what we call cognition (and it is this which participates in the Distributed Cognition discussed earlier). Part of the operation is consciousness – that controlled by attention – and part is not: the subconscious. Psychoanalysis, in this scheme, is about trying to read the inscrutable (and inherently idiosyncratic) off-line component of the model (i.e. the mind) via its projections into the (de-) sequencer. Notice, here, that dreams may be such projections but stripped of attentional management. Not exactly subconscious, but not managed cognition either, a sort of uncontrolled inner experiential world running on the basis of "free association" instead of being within an attentional harness. Where such control, or anchoring in time, is partially present we have day-dreams and the strange states of not being entirely awake or entirely asleep (or the transitions into and out of sleep: hypnagogic and hypnopompic states). Additionally, hypnosis would appear to be the passing of attentional control – the reins of consciousness – via language to another person (and perhaps this is also the case in mass hysteria). The attentional grip on the (de-) sequencer is perhaps much more fragile than we'd like to imagine.

3(9).19.6.4 We can revisit the last two GCPs in the light of the foregoing. These are reproduced below for the convenience of the reader.

GCP7 Cognition and affect can be distributed and projected:
- i) in the environment – space, objects, others (not just con-specifics),
- ii) in time – historical, personal.

GCP8 Thought is the production of cognitive entities.

3(9).19.6.5 GCP7 is really about the operation of the (de-) sequencer. Aspects of behaviour – structure and attention – can be externalised along with the behaviour itself. A person witnessing the behaviour of another does more than just encounter the fine scale output entities. Larger scale entities are recognised and responded to. This is particularly obvious in speech – the output sounds of the speaker have to be processed and contextualized, along with the syllables and words, the morphemes and stresses, and etc., up to the intentions of the speaker. Likewise, with tasks it is the case that a witness appreciates both the task and the component activities required to achieve it. Indeed – unwelcome advice from onlookers is frequently offered, reflecting the readiness with which people assess witnessed activity for any match between perceived goal and perceived actions in relation to achieving that goal. In the case of affect – an attentional mechanism – the same runs through; witnesses go beyond the immediate display of affect, and the final fine scale output behaviour entities, to appreciate the attentional work being done. The efforts can sometimes be on a grand scale (ceremonies) but do not have to be. Projection, as opposed to distribution, requires only that the originator is able to recover the links or associations made. The "output" is not for public consumption, so to speak, but may aid the projector in making public other aspects of cognition and affect. Both distribution and projection do not have to be part of consciousness – but they are part of the functioning of the (de-) sequencer and are thus part of cognition in the general sense (see also 3(9).18.1.3).

3(9).19.6.6 The above account works through, *mutatis mutandis*, for other species. There can be co-operativity in hunting, or in tending to nestlings, or whatever. Of course, there may be beings (species) in which isolated actions/perceptions are the norm – but even these have to procreate, at the very least, or engage in other activities which show awareness of conspecifics and predators. And exploitation of context is going to be part of the behavioural repertoire. Further, consciousness, as the experience of (part of) the mind-body solution, will in some measure (to be determined) be available in non-human species.

3(9).19.6.7 GCP8 presents the idea that thinking is not about consciousness or the internal speech stream. The important point is that entities have to be created and that this could happen off-line in the mind, or via the (de-) sequencer perhaps when learning something or seeing something new, or perhaps as a consequence of internal speech (but not "in" internal speech). The concern with the formulation of the GCP8 is that thinking is not self-evident and is not "done" in any medium. We do not "think in speech" or "think in language" for example. Thoughts, plans, intentions, etc. may be "revealed" to consciousness when they are projected into the (de-) sequencer (or not, as in the car keys example above), they may be created via the (de-) sequencer, they may be perceptual novelties requiring the construction of new cognitive entities for the viewer. Note, in the general case there is nothing specifically human about thinking; the involvement of inner speech is human, but is not required for thought.

3(9).19.6.8 The reader should not assume that the "cycling" of output from the BGP, via injection into the (de-) sequencer and then "reading" back into the BGP is only done with language – one's inner speech. Just as with attention management for control of the Sequential Imperative, and for exactly the same reasons, such thinking can feel as if it is being done "in" different (non-speech) behaviour, such as rehearsal of assembly of components of a mechanism, or the layout of plants in a garden. Thinking in the BGP (creation of cognitive entities) is offline, but can become the topic of conscious attention/ focus through injection into the (de-) sequencer for subsequent "re-perception". This is yet another aspect of the way that the BGPs are not static; but note again, the activity in the BGP is not entrained by the flow of time in the external world.

3(9).19.7 In summary thus far, then, the sensorium is irrevocably idiosyncratic and personal, and not shareable/ knowable beyond the generality of its contribution to one's mental world. The fact of this idiosyncracy is unproblematic – synaesthetes are sometimes unaware that their perceptual experience is unusual, and the fact of synaesthesia does not noticeably impinge on daily life (apart, perhaps, for some a certain quickness at spotting typographical errors). As a human, but also as an animal with a brain, I can in some sense "know what it its like" to have a sensorium. Avian vision, as noted earlier, can be so different from our own that this "knowing" is really rather limited. And of course, we cannot know that the green we see in our head when looking at green grass is the same green experienced by another person. Notice, incidentally, we cannot choose to see the grass as blue.

3(9).19.8 The mind is the off-line processing and storage, as already discussed, which projects entities into the (de-) sequencer and into which the (de-) sequencer injects data – it is idiosyncratic, personal, but also inscrutable, known only via activity in the (de-) sequencer. The functioning or operation of the (de-) sequencer is cognition, and some of this is controlled by attentional behaviour (also projected from the off-line processors) giving the multifaceted experience of attention which we call consciousness. Language is an attentional mechanism, it was proposed in the previous chapter, and so "internal speech" as an attentional mechanism, complete with sensations, can be part of the experience we call consciousness. There can be other means by which the involvement of the (de-) sequencer contributes to thinking; one can rehearse assembly of items, or movements of parts/ limbs... all of which can feel like "thinking in" some sort of behaviour. But just as with "inner speech" the behaviour is not a medium for thought (see 3(9).20.4.2). Note that consciousness has no function, just as the colour red, or the timbre of a French horn, have no function.

3(9).20 So what, finally, is it like to be an ETI? The reader, by this point, should feel comfortable with the idea that the Sequential Imperative is really universal – where there is a brain so must there be the SI. Note that Nagel (1974) implied that his concern for the mind-body problem, consciousness, and etc., reflected a concern for universal properties of beings, including Martians. And more recently (Nagel, 2011), he elaborated the notion that post-material explanations of organisms with life require some sort of account of what might very crudely be called "mental life force" (although of course he doesn't use that term) which is universal in the same way that our understanding of the physics of atoms is universal (but without being reducible to such a physical account). We will return below to look again at the mental world of humans; ETI is a useful thought experiment, as much as anything.

The functional specification of the brain – to serve the SI – appears necessarily to require certain functional components. The (de-) sequencer is required irrespective of whether or not one finds attractive the spiral bound notebook model with half-planes and a data structure called the spine. The notion that some part of the brain runs processes and storage which are offline seems uncontentious – how else could the SI operate? Likewise, management of the SI is always going to be required. All of which means that knowing what it is like to be human, which as a human one does, of course, despite the earlier concerns, is going to be not far off knowing what it is like to be an ETI. Sure, there will be huge differences of detail – but the biology may not be so different (weird life is probably science fiction – see Baross, 2007; Sullivan and Baross, 2007; and for detail see Lane, 2015). Sexual reproduction is required for evolution (I'm assuming it is the major source of genetic change), and in the case of our strongly altricial style of infant development – with long periods of care from the parents – evolution has been constrained by birthing geometry. This may not apply to ETI of course, but if ETI is born without the equivalent of our neurological plasticity or long developmental support from parents then socio-cultural development will be very different (see 3(9).18.3). However, before we delve too deeply into the mental world of ETI, what can we say about the physics and biology?

3(9).20.1 In 1955 Eric Frank Russell published a science fiction short story entitled *The Waitabits*. This amusing story invites the reader to imagine a world where everything to do with life happens excruciatingly slowly. There are some necessary gaps in the account to make the fiction flow – natural phenomena are not discussed, for example. Railway trains move at snail-pace, as do motor-bikes and the *Waitabits* themselves when on foot. The point of the story is that life at such a slow pace is literally not commensurate with human experience and thus the *Waitabits* are essentially intelligent but unknowable. The significance of the *Waitabits* for our exploration here is that they raise the question of the pace of sequentialization, irrespective of its organization. And the answer is found in the missing element in the story – natural phenomena. If one of the *Waitabits* were to trip, for example, and fall over, they would fall to the ground pretty much as fast as here on Earth (the mass of their planet – Eterna – was given as 110% the mass of the Earth). Rain would not happen in slow motion, the effect of objects rolling down hill, or colliding and demonstrating causality, would all happen at "normal pace" so to speak (and Newton's laws of motion would not be changed). Slow biological motion, as imagined, would make no sense (and would surely not have evolved – one is asked to believe it might have done because their day and night were each equivalent to 6 months on Earth). But the pace of the natural world, of chemical reactions, of motions of prey and predator, would all be of a piece. To be sure there would be variations, as we see here on Earth, but within a limited range. All of which is to say that there is inevitably a link between the rate at which events happen in the "real world" and the rate at which entities pass in and out of the body through the skin. Of course, if the body is a bird of prey, or a bat, the behaviour may happen so fast as to leave mere humans gasping with astonishment. Humans are not privileged – but the overall scale of activities is always going to be tied to events in the world. And for ETI this will be the case as well.

3(9).20.2 Can we say anything about ETI's world and biology? The current astrobiology presumption is that life is water-based, with carbon as the basis for organic chemistry. Weird life scenarios based on alternative biochemistry seem unfeasibly weird (Baross, 2007; Sullivan and Baross, 2007; Lane, 2015), but some people will always note that so little is known, really, that nothing is impossible. For example, there is some weirdness in life found on earth. In very cold aquatic environments we can encounter here on earth copper based oxygenation systems in blood, but they are not as efficient as the more widespread use of iron in haemoglobin. Likewise, there has been a study reporting that arsenic might work in place of phosphorous in some organic chemistry (Wolfe-Simon, et al., 2011) but the work has been challenged (and see Drahl, 2012). Given the absence of much real life weirdness, and the lack of evidence for life 2.0 here on earth (see Davies, 2010 and Lane, 2015), it seems prudent to assume that water-based life is the norm, and that therefore other planets bearing life – intelligent or otherwise – must be earth-like in many ways. The search is thus on for earth-sized planets in the "Goldilocks Zone" around stars with appropriate spectral characteristics, where the planets have liquid water and oxygen in sufficient abundance for life as we know it to evolve. Life a bit like, but not quite exactly the same as, life as we know it can be envisaged (perhaps involving different chemical isomers); but life forms – of course – are another matter. Evolution is assumed to be driven ultimately by energy consumption – speciation defines ecological niches for energy exploitation, and creatures evolve by escaping one niche and defining another through differential expression of the Sequential Imperative. Intelligence, in this scheme, is merely a way to boost exploitation of energy, and the SI provides for speciation.

3(9).20.3 ETI – assumed to be intelligent enough to search for us if it can afford to (it might try building humongous orbiting optical telescopes) – will have to have evolved with sufficient sensory acuity to know about stars and astrophysics and all the rest of it. Their vision system might be more like a terrestrial bird's, perhaps, but it will have to be able to see in some way or another. Their brains will have to serve the Sequential Imperative – this is unavoidable. They will rely on sexual reproduction to provide the basis for evolution, but they may have "topped out" on evolutionary development as, perhaps, humans have done (3(9).18.1.2). It is not sensible to suppose they will possess faculties we cannot conceive – for example, telepathy (a staple in much of the Science Fiction literature) is simply not possible, anywhere, because that is not part of the functional specification of any brain, anywhere, and because each being's internal world is idiosyncratically configured on the basis of necessarily idiosyncratic experience (identical twins notwithstanding). Consciousness in ETI as in humans (and terrestrial animals) will be the experience of the (attentionally managed; 3(9).19.8) operation of the (de-) sequencer.

In short – it's entirely possible that being an ETI is rather like being a human. Will they have polluted their oceans with plastic bags, or wrought destruction through over-enthusiastic development and use of hydro-carbons and nuclear weapons? Quite possibly. Will they have evolved wings or the ability to eat and digest old electric motors? Probably not. The likelihood that ETI will *closely* resemble *Homo sapiens sapiens* is remote. Being constrained to evolve in ways which provide brains serving the Sequential Imperative does not constrain beings to resemble closely those we know. However, evolving a set of beings with brains, and various niches to define/ fill, and etc. – in short, a biosphere – seems likely to produce a recognizable equivalent of an intelligent bipedal primate with good hands, capable of doing astrophysics, biochemistry and all the rest. Humans could probably engage in Distributed

Cognition and shared tasks with ETI. Will ETI have language capabilities and will these have recognizable equivalents of syllables? Quite probably, and for the reasons outlined in 3(8).16.1 in respect of sign language syllables. Will they make tea in various ways – as we humans do? No – they won't have *Camellia sinensis* but they might infuse "leaves" from one of their "plants" to make a beverage. And of course they might do this in a variety of ways, just as humans do. In the sense that ETI must be a being constrained by the Sequential Imperative, with all that we've seen that to mean, I think I know something of what it is like to be an ETI.

3(9).20.4 There remain a couple of rather more challenging issues which require more to be said about humans and their brains. But first, a summary: The mind, in the model developed in this book, is the inscrutable background storage, processors and processing activity which projects into the (de-) sequencer material for sequencing, and which in reverse receives from the same (de-) sequencer material for storage and processing. The (de-) sequencer operates bi-directionally so that simultaneous "top-down" and "bottom-up" processing of activity and sensory stimulation can be co-ordinated. Where the work of the (de-) sequencer is managed by attentional mechanisms such as language then the sensation of "now", at various timescales, gives rise to the experience of consciousness. Species differences in the experience and expression of the Sequential Imperative will exist, as well as individual differences (lack of homogeneity) within any species.

3(9).20.4.1 The operation of the (de-) sequencer is also what we think of as cognition and where this is not experienced as "now" at any scale then it is subconscious. Thus the operation of the (de-) sequencer is always cognition, but only partly conscious. Which leaves to be explained the notion of a medium of thought, such as language. Thought, in the model, is the inscrutable processing in the background – the mind. Material is projected from the mind into the (de-) sequencer to contribute to the on-going cognition, and where it may become experienced as part of "now" and thus conscious. If the material is injected as if for linguistic externalization then we have the experience of "inner speech". This can, of course, be de-sequenced for the background processing (re-injected into those processes) but because both the sequencing and de-sequencing are complex processes in the bi-directional operation of the (de-) sequencer the "inner speech" is not just a report of thought, but a *contribution* to the background processing through its influence on that processing. The "inner speech" or "internal language" is not a *medium* for thought.

3(9).20.4.2 Thinking is not, therefore, done "in" language/ speech, as it seems to us when conscious, but is accompanied by speech (or indeed other behaviours – more on this in a moment) and can thereby be entrained in the "now" which permits thinking to be directed by attention and coupled into experience and context, shared with others, and so forth. It doesn't have to be, but can be. The other behaviours could for example be such activities as visual rehearsing of the execution of a carpentry task – envisaging the cutting, shaping, arrangement, fixing, and fastening of bits and pieces in order to assemble a worktop or shelf, for example. Thinking is not being done "in" anything. The carpenter is scrutinizing the space and "thinking through" the sequences and requirements, probably with visual imagery, and in the context of known resources, tools and skills. The point is that the experience is in the "now" and thus conscious. There may be personal preferences for making thought conscious, and for running the selected activity through the (de-) sequencer to bring the thinking into conscious experience and open to attentional direction. This does not amount to "thinking in" anything, and especially not language.

3(9).20.4.3 Is creativity is a challenge for the functional model of the brain? In the account presented in this book creativity is part of thought and thus can result in the projection into the (de-) sequencer of material for possible externalization, re-injection back into the background processes (BGP), and so forth. Creation of text, of artefacts, of music, etc., happens this way. But creation is more than just production – it is also found in perception. The cover image chosen for this book[6] can be seen as a face: grey hair, pink face with brown eyes, a rather straight nose, and an odd mix of discoloured teeth and beard. When I framed the image in the camera I specifically had that sort of interpretation in mind – that portion of the façade of the long building looked to me like a face. The interpretation of the image as a face is clearly the result of some "top-down" processing interacting with the "bottom-up" processing to yield an amalgam of different structural interpretations. But as this processing cycles through the (de-) sequencer and the BGP it sustains a more generic account whereby buildings can be seen as faces, windows as eyes, and so forth. Indeed, such interpretations may have already been encountered on other occasions and are here being applied to this specific image. And for each and every viewer of the image the creative act of "seeing a face" will be different, and in some cases will be absent.

6 Photographed on Suomenlinna. The long building, of which the image shows just a couple of windows, houses a library: Suomenlinnan kirjasto.

In relation to music, the composer Sofia Gubaidulina has referred to the transformation between "vertical time" (in the mind) and "horizontal time" (the sound sequence). In an interview with Andrew Clark, reported in the Financial Times,[7] there is this passage:

> Gubaidulina recalls Schnittke telling her that his inspiration invariably came from a single musical cell, which he developed slowly and painfully until all was revealed. "In my case it's the opposite. In the first instance, perhaps on a walk, I hear a huge, shapeless, multi-faceted sound, absolutely fascinating, with everything piled up together in a way you could never notate – something which exists outside time. It's like a present, and I consider it a duty to transform it from vertical to horizontal."
>
> "Why do I compose? Because I can't leave this fantastic sound hovering in mid-air. It's a big task to unravel it. I sense a moment here, a moment there, then suddenly I come to the beginning, and that's when I start to write it out, without quite knowing how it will develop."
>
> She says audiences experience the process in reverse order. "Thanks to the hidden power of memory, the listener follows the horizontal form, and builds it back into the vertical."

Gubaidulina's account of the process within the listener is very much the same as the conception in this book of de-sequencing in the "curious sort of bi-directional time machine" mentioned at the start. Indeed, her account both of her own creative process and of Schnittke's seems like an attempt to describe the projection into the (de-) sequencer of material in the mind/ BGPs.

3(9).20.4.4 All of which is to say – creativity must depend upon the Sequential Imperative to get into and out of the mind, and it must do this as much as any other activity. What goes on in the mind, the BGP, "off-line", or however else one might describe the inscrutable part of one's mental activity (and it is activity), is really only known to us in a derivative way following traffic through the (de-) sequencer. In that sense creativity is no different from other behavioural intentions. Creative artefacts, or "outputs" are Projected Cognition which morphs into Distributed Cognition in, or with, those who respond (creatively) and reveal understanding of the creator's intention. And thus this book too is the

7 Andrew Clark. (*Financial Times,* 1999, Jan. 11, p. 18). "The creative eloquence of silence: Andrew Clark meets composer Sofia Gubaidulina on the eve of her visit to London with her new work". The journalist Andrew Clark is not the same person as the philosopher Andy Clark, referenced earlier in connection with cybernetics (2(6).11.6).

output of my (de-) sequencer's processing of inscrutable thoughts projected into it; checked back to it via perception and then re-sequenced with changes (we call this revision) and back and forth (more revision) until a final read-through matches well enough my intention. But always with the possibility of further revision! And for the reader a similarly creative cycling process takes place whereby the "content" of the book comes to be understood. One thing wrong with Neisser's account of perception (1(1).4.1) is that it is too simple and uni-directional – it served the purpose at the start of our journey, but can be seen to be inadequate at journey's end.

3(9).20.4.5 What about the notion that language is actually about communication, and maybe has just been pressed into service (for managing the Sequential Imperative) because it is already in place in humans? This isolates humans as exceptional and thus the rest of the animal kingdom as requiring a different account. But, what actually is meant here by communication? Clearly hiding in some bushes and watching yesterday's wounded beast limp by on its way to the waterhole provides an opportunity for linguistic interaction which is very context-bound. This is a world away from writing this book, or listening to a lecture on 19th Century French oil paintings or 20th Century nuclear power station design. Surely? And, isn't it "obvious" that we communicate the thoughts we have framed in inner speech?

If instead you reconceive communication as being about ensuring that your interlocutor can de-sequence your sequencings then the rôle of the Sequential Imperative becomes clearer – it has to be served, and communication is about making that possible in an ever wider range of circumstances by extending the "stuff" available to the background processes. The problem addressed by communication is how to work with heterogeneity of experience, memory and understanding when (de-) sequencing and managing the attentional focus of the people concerned. It seems to me, at least, that this is actually no different from managing attention for multi-tasking. Colloquially, it is about being on the same page; it is about being understood well enough to move on from discourse repair.

3(9).20.4.6 Which leads me to one final thought. Aside from individual differences and idiosyncratic mental worlds what else can we say about minds in society? It would seem to me that each and every one of us functions in a mental world in which there is no idle capacity. We are all "cognizing to the max" (i.e. a personal max), so to speak, but this casual remark does suggest further work to explore/ analyse the extent to which it is or is not a sensible summary of how the brain delivers on the SI. What may be tragic is that through circumstance or misfortune some of us may not get the chance to develop our (de-) sequencer and BGP as fully as we might have done (perhaps through insufficient stimulation in critical periods). What may be exciting is that studies of cognitive malfunction may be re-energized by being cast in terms of the Sequential Imperative and a brain's inability to serve this in one or more ways. This is not fanciful – Lieberman (2006), although motivated to look toward the possibility of an evolutionary biology of language, discusses cognitive malfunctioning consequent upon the effects of hypoxia on the basal ganglia – his "sequencing engine".

3(9).20.5 So we come full circle – and return to the relationship between the functional specification and the neurophysiology. The functional specification of the brain – any brain, any species, anywhere, any time – does not change with biology. But within a species, or life form, it may transpire that mapping the functional specification, with all its details, onto the biology permits biological faults to be understood in functional terms, and vice versa. So if a time-switch mechanism is made of brass, rather than plastic, we may come to expect certain types of fault in similar devices; plastic would yield a different spectrum of faults. This doesn't break the focus of attention on the functionality. Rather, it demonstrates that only when the functionality is fully understood can it ever make sense to consider the ways in which the functionality is delivered. And so with Aliens – different life forms, different biologies, different neurophysiologies, ... but all constrained by the Sequential Imperative and the intricacies of its management.

Afterword

In the Preface I noted "that the book is explicitly a prolegomena – a set of notes for bigger projects to come. It is a monograph not a textbook; it is broad in scope but without a monumental list of references or doorstop heftiness. The nature of notes is incompleteness and, to switch to a graphic metaphor, the set of notes amounts to a sketch which can do much to reveal the overall picture, along with the size of the canvas, without covering it all. Some material will be well detailed, some in outline awaiting adumbration, and so forth. But all combine to depict a totality in the image." I believe the book has delivered on this promise. It is now clear enough what the sketch has covered – and sometimes in significant detail, as in the discussions of speech production.

All beings with brains are forced into sequential activity, and their brains manage this Sequential Imperative. It all sounds so simple: Perception is de-sequencing, behaviour is sequencing. The brain stores "stuff" and derives cognitive "stuff" from perception and produces behavioural components for sequencing. But it all turns out to be more complex in the details. One such detail is that the "stuff" is not entirely atemporal as characterised in the opening Part of the book: learning and thinking require some activity in the background processes. This activity is not, however, entrained by or aligned with external/real time; the curious sort of time machine is still in place and still with work to do! One particular detail is that the management of the (de-) sequencer turns out to be crucial, and it is for this, perhaps, that language emerged in humans. Likewise, it turns out that a few General Cognitive Principles are necessary to turn the simple idea into a deliverable scheme for exploring the functionality of the brain.

Primates other than humans may have evolved their (de-) sequencer complexity to something close to ours – but sufficient for defining their ecological niches. Perhaps more visual processing for gloomy evenings – but at the expense of other activities. Perhaps with enough of a sense of "now" to be able to plan the utilisation of fire from one day to the next, or to identify stone cores for eventual use in tool making. Perhaps to enable vocalisations and grooming, but not discussion of the weather or the need to move to a new location. These are not steps on the inevitable road to the emergence of humans; simple linear notions of evolution as progression make no sense (and see Tattersall, 2012). It is possible that the emergence of symbol use in the archaeological record flags the evolution of enough Sequential Imperative resource to permit the individuals concerned a longer "now" and more complex attentional work which could support something like proto-language; but this leads us to perhaps the biggest project which might be follow-up work to this book (see also 3(8).17.6).

© KONINKLIJKE BRILL NV, LEIDEN, 2017 | DOI 10.1163/9789004342996_015

For example, the idea that the emergence of symbol use in the record flags the emergence of a new language-capable brain (and thus *Homo sapiens sapiens*) is not obviously correct. It would seem more likely that our species emerged some time before we see evidence of symbol use. The reasoning here is that what emerged in *Homo sapiens sapiens* was the cognitive capacity to manage the environment more efficiently/ successfully; in other words more complex multi-tasking. Raising children who are so dependent during the first years of life requires reliable supplies of food and support. Group size might be quite large but efficiently fed. Only when patterns for such living are established can the multi-tasking capability be exploited in ways not so immediately linked to providing the next meal. That is when the symbols enter the record – the opportunities exist, and the skills and capabilities also exist. We know from the emergence of writing that something similar happened in the Bronze age – sufficient resource and tool-making prowess made living even easier. So devoting time to specialist activities such as writing became easier. Notice that the idea of writing spread rapidly, even if the systems did not (see also 2(4).8.2). We don't assume the emergence of writing was due to some change in the brains of *Homo sapiens sapiens* around the world – so why should we assume that the first emergence of symbol use in the record flags such an event?

The bigger projects identified in the book for following up include the need to assess alternative sets of General Cognitive Principles. The Sequential Imperative is unavoidable, but that does not mean that the GCPs offered here, and the overall account of how the brain supports or delivers the SI, are equally unavoidable. Additionally, there is a need for a comprehensive account of language in whatever overall account is chosen; linguistic theory needs working on beyond the sketched non-linear structure offered here (that Cognitive Science, as presented here, provides for the possibility of language seems clear – but it does not amount to a linguistic theory). Work on computational modelling of the delivery of the functionality also needs to be done – this could be a major new component of work in Artificial Intelligence. The implied evolutionary account of brains includes the notion that as multitasking emerged as useful so it became necessary to manage the multi-tasking, and for this language (itself dependent on, and demonstrating, extensive multi-tasking) became valuable. The underlying notion is that the elaboration/ increase in multi-tasking capability is capacity limited by management requirements (such as language). Essentially the evolution of complexity in the management of the Sequential Imperative (intelligence) "tops out" (see 3(9).18.1.2). This bigger project needs more work. And, of course, claims are made about consciousness and the mind-body problem, and these need to be worked up in greater detail.

Homeostasis can be anchored in the functional specification of whatever sort of brain one can find in very small creatures. Evolution of the brain to manage ever more complex homeostatic requirements may have been the early driver for increased complexity and speciation. This is another bigger project for the future. It is interesting that when the Sequential Imperative is explained to people, for example in a conference presentation, someone will ask "how small can the creature be and still conform?". The acceptance of the notion of scale in serving the Sequential Imperative is crucial to understanding the evolution of the brain, and of the emergence of apparent wonders such as language. No specific saltation is required.

The account offered in this book licenses serious reflection on the capabilities of Extra-Terrestrial Intelligences. Will ETI have music and cave drawings and all the rest? What about ETI's semiosphere? I think it likely that ETI will have something we could call a semiosphere – the historical/ cultural/ social mental milieu within which we all exist – sometimes different for different groups of people, sometimes much more widely shared. Following reflection on ETI can we support arguments that evolution of intelligence inevitably "tops out", as mentioned above and earlier in the text? Another big project. And related to this is the need to develop a thorough account of creativity – yet more big project work!

And what then of non-humans here on earth? The Sequential Imperative is unavoidable; some animals will have complex ways of managing and exploiting the serving of the SI, others less so (compare/ contrast dolphins, crows, cats, dogs, bonobos). Another big project is work on the details of behaviour in non-humans – much is assumed here but these assumptions amount to a prediction that lots of details will indeed be uncovered. This book seems to have ended up being about human language, thought and mind, but we should not lose sight of the generality of the story: the inevitability and unavoidability of the brain – any brain, any species, anywhere, at any time – serving the Sequential Imperative. *That* is what the brain is for.

"Yes, but where is the evidence?" I hear the reader exclaim. To which the answer goes: If the story told here is plausible, perhaps even blindingly plausible, for human speech, then it must run through for other behaviour in humans and also non-humans because otherwise one is left with an endless requirement for special stories to be told about behaviour in each and every other species – *yet they all have brains*. Furthermore, why the reluctance in respect of the brain when not in respect of the eyes, or mouth, or limbs, or heart, or…? Do we really believe we can work with generalized accounts of organs, articulators and sensors – but not the brain? So the data on other species remain to be uncovered in detail but the conjecture (is hypothesis too strong a word?) offered in this book must, at the very least, tell us what we should be looking for, as well as how to make sense of it all.

But then maybe you don't "get" the idea of the Sequential Imperative at all. In that case the challenge is for you explain to us all what the brain does in all species, and how your account makes possible human language. This last point is non-trivial. The SI is a functional account of the brain. As noted earlier – if a special case is claimed for language then that cannot replace what the SI (and serving it) does accomplish in respect of language. The theorist is not free to say that the SI does not operate in any specific domain. Special accounts of specific behaviour have to be set within the working of the SI. Likewise, for consciousness, the functional account offered here has it that the operation of the (de-) sequencer is, in part, experienced as consciousness (attention drives the experience – there is too much going on in the (de-) sequencer for it all to be experienced). If you seek to argue that consciousness is something else then you need to explain how it fits functionally, and also what the experience of the (de-) sequencer actually is. Note that in both cases – language (for controlling attention) and consciousness – the account is not restricted to humans. The functional account scales with sensorimotor complexity and brain size – but the principle remains. Alternative accounts have to address such issues.

The reader may wonder, before closing the covers of this book, if I perpetrated some sleight of pen when at the start I suggested that we need not look at modern discussions of Cognitive Science because we aimed to reach a different understanding of Cognitive Science. I encouraged the reader to think that we could envisage the Cognitive Scientists of 60 years ago taking a different route had they attended selectively to material other than that they did fasten on to. But then, by the end of the book the work being cited is not just mid 20th Century. The point, of course, is that by now we have so much more work to look at that I can selectively pick out illustrations for my account; this book is not a survey. I believe I can claim that Cognitive Science is now about the Sequential Imperative and the functional specification of the brain – any brain, anywhere, at any time, in any species.

References

Abrahamsen, A. and Bechtel, W. 2012. History and Core Themes. Chapter 1 of Frankish and Ramsey, 2012.

Aikhenvald, A.Y. 2003. *Classifiers: A Typology of Noun Categorization Devices.* Oxford University Press.

Anderson, S.R. 1985. *Phonology in the Twentieth Century.* The University of Chicago Press.

Armstrong, D.F., Stokoe, W.C. and Wilcox, S.E. 1995. *Gesture and the Nature of Language.* Cambridge University Press.

Baross, J.A. 2007. *The Limits of Organic Life in Planetary Systems.* National Research Council of the National Academies.

Bellugi, U. and Studdert-Kennedy, M. (Editors) 1980. *Signed and Spoken Language: Biological Constraints on Linguistic Form.* Report of Dahlem Workshop on Sign Language and Spoken Language: Biological Contraints on Linguistic Form, Berlin 1980, March 24–28. Published by Verlag Chemie, Basel.

Berko, J. 1958. The Child's Learning of English Morphology. *Word* 14: 150–177.

Berwick, R.C. 1998. Language evolution and the Minimalist Program: the origins of syntax. Chapter 20 in *Approaches to the Evolution of Language: Social and Cognitive Bases.* Edited by James R. Hurford, Michael Studdert-Kennedy and Chris Knight. Cambridge University Press. 1998.

Berwick, R.C. and Chomsky, N. 2016. *Why Only Us: Language and Evolution.* MIT Press.

Bickerton, D. 1990. *Language and Species.* The University of Chicago Press.

Bickerton, D. 1998. Catastrophic evolution: the case for a single step from protolanguage. Chapter 21 in *Approaches to the Evolution of Language: Social and Cognitive Bases,* edited by James R. Hurford, Michael Studdert-Kennedy and Chris Knight. Cambridge University Press. 1998.

Birkhead, T. 2012. *Bird Sense: What it's like to be a bird.* Bloomsbury, London.

Breidbach, O. 2006. *Visions of Nature: The art and science of Ernst Haeckel.* Translated by Paul Aston. Prestel Verlag, Munich.

Carruthers, P. and Smith, P.K. (Editors). 1996. *Theories of theories of mind.* Cambridge University Press.

Carstairs-McCarthy, A. 1999. *The Origins of Language: An Inquiry into the Evolutionary Beginnings of Sentences, Syllables and Truth.* Oxford University Press.

Cherry, E.C., Halle, M. and Jakobson, R. 1953. Toward the Logical Description of Languages in Their Phonemic Aspect. *Language*, Vol. 29(1).

Chomsky, N. 1980. *Rules and Representations.* Basil Blackwell, Oxford.

Chomsky, N. and Otero, C-P. 2004. *Language and Politics.* 2nd, expanded, edition. AK Press, Oakland and Edinburgh. First edition: 1988.

Clark, A. 2013. Whatever next? Predictive brains, situated agents, and the future of cognitive science. *Brain and Behavioural Sciences,* 36: 181–253.

Cline, E.H. 2014. *1177 B.C.: The Year Civilization Collapsed.* Princeton University Press.

Coleman, J. 1992. The phonetic interpretation of headed phonological structures containing overlapping constitutents. *Phonology,* 9(1): 1–44.

Comrie, B. 1989. *Language Universals and Linguistic Typology.* 2nd edition. Blackwell.

Craig, C. (Editor). 1986 *Noun Classes and Categorization.* John Benjamins.

Croft, W. 1990. *Typology and universals.* Cambridge University Press.

Damasio, A. and Carvalho, G.B. 2013. The nature of feelings: evolutionary and neurobiological origins. *Nature Reviews: Neuroscience.* 14: 143–152.

Davies, P. 2010. *The Eerie Silence: Are We Alone in the Universe.* Allen Lane.

de Léon, D. 1999. Building Thought into Things. *Proceedings of the 3rd. European Conference on Cognitive Science,* pp. 37–47. Siena, Italy.

de Wall, F. 2016. *Are we smart enough to know how smart animals are?* W.W. Norton and Co., London and New York.

Diringer, D. 1968. *The Alphabet: A Key to the History of Mankind.* 3rd edition. Hutchinson.

Drahl, C. 2012. Two Studies Rebut Arsenic-Based Life. *Chemical and Engineering News,* 90(29). http://cen.acs.org/articles/90/i29/Two-Studies-Rebut-Arsenic-Based.html.

Edmondson, W.H. 1981. Sign Language in an Unfavourable Setting: A perspective. In *Perspectives on British Sign Language and Deafness.* Edited by B. Woll, J. Kyle, and M. Deuchar. Croom Helm.

Edmondson, W.H. 1986a. Issues in Linearization: Prolegomena for a general theory of communication. *Language and Communication,* 6(4): 225–266.

Edmondson, W.H. 1986b. Autosegmental Phonology: The syllable and the sign. In *Signs of Life.* Edited by B.T. Tervoort. Amsterdam: University of Amsterdam, Institute for General Linguistics, publication 50.

Edmondson, W.H. 1989. A Non-Concatenative Account of Classifier Morphology in Signed and Spoken Languages. In *Current Trends in European Sign Language Research.* Proceedings of the 3rd European Congress on Sign Language Research, Hamburg, July 26–29. Edited by S. Prillwitz and T. Vollhaber. SIGNUM Press. Hamburg.

Edmondson, W.H. 1990. Segments in Signed Languages: Do they exist and does it matter? In *SLR'87: Papers from the Fourth International Symposium on Sign Language Research.* Edited by W.H. Edmondson and F. Karlsson. SIGNUM Press. Hamburg.

Edmondson, W.H. 2002. Synaesthesia and Colour Vision: The Personal is philosophical. Invited paper for *Zeitschrift für Semiotik,* editor R. PosnerProf, T-U Berlin. This paper was invited following my presentations at the conference on Gesture in Portugal in April 2000. Published as: Farbige Buchstaben: Synästhesie und Sprache. *Zeitschrift für Semiotik* 24(1): 51–64.

Edmondson, W.H. 2010. General Cognitive Principles: The Structure of Behaviour and the Sequential Imperative. *International Journal of Mind, Brain & Cognition.* Vol. 1(1), Jan–June 2010.

Edmondson, W.H. and Beale, R. 2008. Projected Cognition – Extending Distributed Cognition for the study of human interaction with computers. *Interacting with Computers*, 20: 128–140.

Edmondson, W.H. and Zhang, L. 2002. "The Use of Syllable Structure for Speech Recognition". University of Birmingham. School of Computer Science. Technical Report CSRP-02-7, June 2002.

Feinberg, T.E. and Mallett, J.M. 2016. *The Ancient Origins of Consciousness.* MIT Press.

Firth, J.R. 1957. *Papers in Linguistics: 1934–1951.* Oxford University Press.

Fodor, J.A. 1983. *The Modularity of Mind: An essay of faculty psychology.* MIT Press.

Frankish, K. and Ramsey, W.R. (Editors) 2012. *The Cambridge Handbook of Cognitive Science.* Cambridge University Press.

Freud, S. 1901. *Psychopathology of Everyday Life.* Translated by A.A. Brill, Published by Macmillan, 1914.

Friedman, L.A. (Editor). 1977. *On the Other Hand: New Perspectives on American Sign Language.* Academic Press.

Fox, M. 2013. *The Riddle of the Labyrinth.* Profile Books, London.

Geldard, F.A. 1953. *The Human Senses.* John Wiley & Sons, New York.

Goldsmith, J.A. 1976. *Autosegmental Phonology.* Doctoral Dissertation, MIT. Garland Press, New York. 1979.

Goldsmith, J.A. 1990. *Autosegmental and Metrical Phonology.* Blackwell.

Goldsmith, J.A. (Editor). 1995. *The Handbook of Phonological Theory.* Blackwell.

Gould, J.L. and Gould, C.G. 1995. *The Honey Bee.* Scientific American Library.

Gould, S.J. and Lewontin, R. 1979. The spandrels of San Marco and the Panglossian paradigm: a critique of the adaptationist programme. *Proc. R. Soc. Lond. B* 205: 581–598.

Greenberg, J.H., (Editor). 1966. *Universals of Language.* MIT Press.

Hauser, M.D. 1996. *The Evolution of Communication.* MIT Press.

Hawkins, J.A. 1988. *Explaining Language Universals.* Blackwell.

Hayes, B. 1990. Diphthongisation and coindexing. *Phonology*, 7(1): 31–71.

Hayes, B. 2009. *Introductory Phonology.* Wiley-Blackwell.

Hewes, G.W. 1978. The Phylogeny of Sign Language. In *Sign Language of the Deaf.* Edited by I.M. Schlesinger and L. Namir. Academic Press.

Hockett, C.F. 1960. The Origin of Speech. *Scientific American.* 203: 88–96. Reprinted in Wang, W. S-Y. 1982. *Human Communication: Language and its Pyschobiological Bases.* Scientific American.

Hutchins, E. 1995. *Cognition in the Wild.* MIT Press.

Imai M. and Kita S. 2014. The sound symbolism bootstrapping hypothesis for language acquisition and language evolution. *Phil. Trans. R. Soc. B* 369: 20130298.

Iles, J.P. 1995. *Text-to-Speech Conversion Using Feature-Based Formant Synthesis in a Non-Linear Framework*. PhD. School of Computer Science. University of Birmingham.

Jackendoff, R. 2002. *Foundations of Language: Brain, Meaning, Grammar, Evolution*. Oxford University Press.

Jackendoff, R. 2012. Language. Chapter 9 of Frankish and Ramsay, 2012.

Kahn, D. 1976. *Syllable-based Generalizations in English Phonology*. PhD Thesis, MIT

Kawahara, S. 2016. Japanese has syllables: a reply to Labrune. *Phonology*, 33(1): 169–194

Klima, E. and Bellugi, U. 1979. *The Signs of Language*. Harvard University Press.

Ladefoged, P. and Johnson, K. 2011. *A Course in Phonetics*. Wadsworth.

Lane, N. 2015. *The Vital Question: Why is life the way it is?* Profile Books, London.

Lashley, K.S. 1951. The Problem of Serial Order in Behavior. In: *Cerebral Mechanisms in Behavior. The Hixon Symposium*. Edited by Lloyd A. Jeffress. Republished in facsimile in 1967: Hafner Publishing Company, New York.

Lakoff, G. 1987. *Women, Fire, and Dangerous Things*. Chicago University Press.

Lenneberg, E.H. 1967. *Biological Foundations of Language*. John Wiley.

Levelt, W.J.M. 1993. *Speaking: From intention to articulation*. Paperback edition: Bradford/MIT Press. Hardback edition: 1989.

Levinson, S.C. 2013. Recursion in pragmatics. *Language*, 89: 149–162.

Lieberman, P. 2006. *Toward an Evolutionary Biology of Language*. Belknap/Harvard.

Linde, C. and Labov, W. 1975. Spatial networks as a site for the study of language and thought. *Language*, 51(4): 924–939.

Lotman, Y.M. 1990. *Universe of the Mind – A semiotic Theory of Culture*. Taurus, New York.

MacNeilage, P.F. 2008. *The Origin of Speech*. Oxford University Press.

Mandel, M. 1977. Iconic Devices in American Sign Language. In *On the Other Hand: New Perspectives on American Sign Language*. Edited by L.A. Friedman. Academic Press.

Martin, R.D. 1998. Comparative Aspects of Human Brain Evolution: Scaling, Energy Costs and Confounding Variables. In *The Origin and Diversification of Language*. Editors: N.G. Jablonski and L.C. Aiello. Wattis Symposium Series in Anthropology. Memoirs of the California Academy of Sciences, No. 24. San Francisco, California.

McCarthy, J.J. 1981. A Prosodic Theory of Nonconcatenative Morphology. *Linguistic Inquiry*, 12(3): 373–418.

McClelland, J.L. and Rumelhart, D.E. 1986. *Parallel Distributed Processing: Explorations in the Microstructure of Cognition. Volume 2: Psychological and Biological Models*. MIT Press.

McNeill, D. (Editor). 2000. *Language and Gesture*. Cambridge University Press.

McNeill, D. 1992. *Hand and Mind: What gestures reveal about thought*. Chicago University Press.

McNeill, D. 2012. *How Language Began.* Cambridge University Press.

Meier, R.P., Cormier, K., and Quinto-Pozos, D. 2002. *Modality and structure in signed and spoken languages.* Cambridge University Press.

Miller, G.A., Galanter, E., and Pribram, K.H. 1960. *Plans and the Structure of Behavior.* Holt, Rinehart and Winston.

Nagel, T. 1974. What Is It Like to Be a Bat? *The Philosophical Review,* 83 (4): 435–450.

Nagel, T. 2011. *Mind and Cosmos: Why the Materialist Neo-Darwinian Conception of Nature is Almost Certainly False.* Oxford University Press.

Neisser, U. 1967. *Cognitive Psychology.* Appleton-Century-Crofts, New York.

Nichols, J. 1998. The Origin and Dispersal of Languages. In *The Origin and Diversification of Language.* Editors: N.G. Jablonski and L.C. Aiello. Wattis Symposium Series in Anthropology. Memoirs of the California Academy of Sciences, No. 24. San Francisco, California.

Ohnuma, K., Aoki, K., and Arazawa, T. 1997. Transmission of Tool-making through Verbal and Non-Verbal Communicataion: Preliminary Experiments in Levallois Flake Production. *Anthropological Science,* 105 (3): 159–168.

Pearce, E., Stringer, S. and Dunbar, R.I.M. 2013. New insights into differences in brain organization between Neanderthals and anatomically modern humans. Proceedings of The Royal Society, B. 280: 20130168; DOI: 10.1098/rspb.2013.0168. Published 13 March 2013.

Peters, A.M. 1983. *The Units of Language Acquisition.* Cambridge University Press.

Putnam, H. 1971. The 'Innateness Hypothesis' and Explanatory Models in Linguistics. In *The Philosophy of Language,* edited by J.R. Searle. Oxford University Press.

Rumelhart, D. E. and McClelland, J.L. 1986. *Parallel Distributed Processing: Explorations in the Microstructure of Cognition. Volume 1: Foundations.* MIT Press.

Russell, E.F. 1955. The Waitabits. In *Astounding Science Fiction,* July 1955, ed. John W. Campbell, publ. Street & Smith Publications Ltd. Republished several times, for example in *Best SF Six,* ed. Edmund Crispin, 1966. Faber and Faber.

Sandler, W. 1989. *Phonological Representation of the Sign: Linearity and Nonlinearity in American Sign Language.* Foris Publications; No.32 in Publications in Language Sciences.

Sandler, W. 1993. A sonority cycle in American Sign Language. *Phonology,* 10(2): 243–279.

Savage-Rumbaugh, E.S., Murphy, J., Sevcik, R.A., Brakke, K.E., Williams, S.L., and Rumbaugh, D.M. 1993. *Language Comprehension in Ape and Child.* Monographs of the Society for Research in Child Development No. 233. 58(3–4), 1993.

Schlesinger, H.S. 1978. The Acquisition of Bimodal Language. In *Sign Language of the Deaf.* Edited by I.M. Schlesinger and L. Namir. Academic Press.

Schoenemann, P.T. 2005. Conceptual complexity and the brain: Understanding language origins. In *Language acquisition, change and emergence: Essays in evolutionary linguistics* Edited by W.S.-Y. Wang and J.W. Minett. Hong Kong: City University of Hong Kong Press.

Schoenemann, P.T. 2006. Evolution of the size and functional areas of the human brain. *Annual Review of Anthropology,* 35: 379–406.

Schoenemann, P.T. 2009. Evolution of Brain and Language. *Language Learning.* 59: 162–186.

Siple, P. (Editor) 1978. *Understanding Language through Sign Language Research.* Academic Press.

Solms, M. and Turnbull, O.H. 2011. What is Neuropsychoanalysis? *Neuropsychoanalysis,* 13(2): 133–145.

Stokoe, W.C. 1960. Sign Language Structure: An Outline of the Visual Communication Systems of the American Deaf. *Studies in linguistics: Occasional papers (No. 8).* Buffalo: Dept. of Anthropology and Linguistics, University of Buffalo.

Studdert-Kennedy, M. 1982. On the dissociation of auditory and phonetic perception. In *The Representation of Speech in the Peripheral Auditory System.* Edited by R. Carlson and B. Granström. Elsevier Biomedical Press.

Suchman, L.A. 1987. *Plans and Situated Actions: The problem of human-machine communication.* Cambridge University Press.

Sullivan, W.T., and Baross, J.A. 2007. *Planets and Life.* Cambridge University Press.

Tattersall, I. 2012. *Masters of the Planet: The search for our human origins.* St Martin's Griffin, New York.

Taylor, A.H., Hunt, G.R., Holzhaider, J.C., and Gray, R.D. 2007. Spontaneous Metatool Use by New Caledonian Crows. *Current Biology,* 17: 1504–1507, September 4, 2007.

Tomasello, M. 1990. Cultural transmission in the tool use and communicatory signalling of chimpanzees? In *"Language" and intelligence in monkeys and apes: Comparative developmental perspectives.* Edited by S.T. Parker and K.R. Gibson. Cambridge University Press.

Tomasello, M. 1999. *The Cultural Origins of Human Cognition.* Harvard University Press.

Trask, L.R. 1999. Why should languages have any relatives? In *Nostratic: Examining a Linguistic Macrofamily.* Edited by C. Renfrew and D. Nettle. Published by The McDonald Institute for Archaeological Research, University of Cambridge, UK.

Van Valin, R.D. (Editor) 1993. *Advances in Role and Reference Grammar.* John Benjamins.

Westcott, R.W. (Editor) 1974. *Language Origins.* Linstok Press, Maryland.

Wolfe-Simon, F., Blum, J.S., Kulp, T.R., Gordon, G.W., Hoeft, S.E., Pett-Ridge, J., Stolz, J.F., Webb, S.M., Weber, P.K., Davies, P.C.W., Anbar, A.D., and Ormeland, R.S. A Bacterium That Can Grow by Using Arsenic Instead of Phosphorus. 2011. *Science* 332: 1163.

Index

www.ingramcontent.com/pod-product-compliance
Lightning Source LLC
Chambersburg PA
CBHW031324230926
51848CB00001B/79
9789004342897